ISBN: 978-1-962557-06-1

Building Your Small Business Website: A Step by-Step Guide to Success

Table of Contents

Why Every Small Business Needs a Website

The Digital Shift in Business

In today's world, a business without an online presence is akin to having a storefront in the middle of the desert—no one knows you exist, and even fewer can find you. The shift to digital has become a necessity for businesses of all sizes, but it's especially critical for small businesses looking to compete in a crowded marketplace. Consumers are no longer relying solely on word-of-mouth or local directories to find businesses; instead, they turn to the internet, conducting research and making purchasing decisions based on what they find online.

The question for business owners isn't whether they need a website, but why they need one. And the answer is clear: a website is a crucial part of establishing your brand, reaching your target audience, and converting visitors into customers.

In this chapter, we'll break down the key reasons why having a website is not just beneficial but essential for small businesses. You'll learn how a website can enhance your credibility, extend your reach, and ultimately drive more business your way.

Building Credibility: Your Online Business Card

Think about the first thing people do when they hear about a new business. They Google it. Whether they're looking for more information, reviews, or contact details, people turn to the internet to validate a business before engaging with it. If your business doesn't have a website, it can immediately cast doubt on its legitimacy. After all, why wouldn't a legitimate business have an online presence?

A website serves as your online business card. It's often the first interaction a potential customer will have with your brand, and as the saying goes, "first impressions matter." An outdated or poorly designed website can do more harm than good, but not having one at all is often a deal-breaker. Consumers expect businesses to have a website that provides essential information such as location, services, contact details, and customer testimonials.

Case in Point: Let's say you run a local bakery. A customer hears about your cakes and wants to place an order. They search for your business online, but find no website, only a poorly managed social media page with outdated information. That customer may move on to a competitor that does have a polished, easy-to-navigate website where they can place an order with just a few clicks.

When potential customers find a well-designed, professional website, they are more likely to trust your business. It gives them confidence that your business is real, operational, and competent. If your website reflects the quality and professionalism of your business, it can be the deciding factor in whether someone chooses to engage with you or a competitor.

A Global Audience at Your Fingertips

For small businesses, a website doesn't just offer local exposure; it opens the door to a global audience. The internet has no geographical boundaries, meaning your products or services can reach potential customers far beyond the limits of your physical location. Even if your business operates within a local market, having a website allows you to be discovered by tourists, out-of-town visitors, or even international buyers.

Let's take the example of a small boutique that specializes in handmade jewelry. Without a website, the boutique is limited to foot traffic and maybe some local advertisements. However, with a website, they can sell their products to anyone, anywhere, at any time. E-commerce integration allows small businesses to offer their products online, creating new revenue streams and expanding their customer base.

Example: Businesses in niche markets, such as custom-made crafts or specialty foods, can attract enthusiasts from around the world who are looking for something unique. A website allows these businesses to showcase their offerings and connect with customers far beyond their local area.

A website also makes it easier for your local customers to find you. With more people relying on Google Maps and other online tools for finding businesses, not having a website can make it harder for customers to locate your physical store, particularly in highly competitive markets.

By extending your reach online, a website not only helps you tap into new markets but also serves as a tool to retain and engage your current customer base.

Customer Convenience: 24/7 Accessibility

One of the greatest advantages of having a website is the ability to provide customers with round-the-clock access to your business. In contrast to a physical store that has specific hours of operation, a website never closes. Potential customers can browse your offerings, learn more about your business, and even make purchases at any time of the day or night. This 24/7 accessibility is particularly important in today's fast-paced world, where people often make buying decisions outside of traditional business hours.

Consider how many times you've browsed products or services in the evening after work, or how many purchases you've made online late at night. The convenience factor cannot be overstated. Having a website allows customers to engage with your business when it's most convenient for them, whether that's during their lunch break, while commuting, or after the kids have gone to bed.

A website can offer features such as online booking systems, e-commerce platforms, and contact forms, which make it easier for customers to interact with your business. Instead of waiting for business hours to call and make a booking, a customer can hop onto your website and secure their spot with just a few clicks.

Example: A local spa that allows online appointment bookings through their website is likely to attract more customers than a competitor that only takes phone reservations during business hours.

In essence, a website adds a level of convenience that modern customers expect. It makes it easier for them to choose your business over others, especially when competing businesses don't offer the same level of online access.

Competitive Edge: Keeping Up with the Competition

It's safe to say that most of your competitors already have a website. If you don't have one, you're starting at a disadvantage. In nearly every industry, a business without a website is seen as behind the times or less legitimate. The reality is that potential customers are more likely to choose businesses that have a website over those that don't, simply because it's easier to access information and make informed decisions online.

Without a website, you're essentially handing your competitors the upper hand. By maintaining a professional online presence, you can compete on a level playing field, even if your business is smaller or newer than your rivals. In fact, a website can often be the tool that helps you differentiate yourself from competitors.

Example: If two local restaurants offer similar types of cuisine, the one with a better website is more likely to attract customers. A website that offers an easy-to-read menu, online reservations, and positive reviews can be the deciding factor.

Your website is also a place where you can highlight what makes your business unique. Whether it's through showcasing your products, telling your brand story, or providing customer testimonials, your website allows you to emphasize the qualities that set you apart from your competitors.

A Website Is Essential for Success

In today's digital world, having a website is no longer optional for small businesses—it's essential. From establishing credibility and expanding your reach to providing convenience and staying competitive, a website offers numerous benefits that can help your business grow. As we continue through this book, you'll learn more about how to plan, build, and maintain the perfect website for your small business. With the right strategy and support from partners, you'll be well on your way to online success.

Your Business Needs

Defining Your Website's Purpose

Before jumping into website design or development, it's crucial to take a step back and assess why you need a website in the first place. Many business owners make the mistake of thinking, "I need a website because everyone else has one," without fully understanding how it can specifically benefit their business. This leads to websites that look good but fail to deliver results.

To avoid this common pitfall, you need to define the purpose of your website. Is it primarily for lead generation, to sell products, provide information, or simply establish an online presence? Answering this question will shape how you approach every decision, from design to content creation.

For instance, an e-commerce business will need features like product listings and a shopping cart, while a service-based business may prioritize appointment booking systems and showcasing customer testimonials. By aligning your website's features with your business goals, you can ensure it serves its purpose effectively.

Pro Tip: Start by listing the top three outcomes you want from your website. Whether it's generating leads, increasing sales, or building brand awareness, having these goals in mind will help guide your decisions during the planning phase.

Evaluating Your Business and Goals

Understanding your business inside and out is the foundation for a successful website. Each small business is unique, and your website should reflect that. Before you start building, take time to evaluate your business, its mission, and how the website can serve as a tool to achieve your goals. Key Questions to Consider:

What does your business do, and how can your website help communicate that clearly to visitors?

Who is your target audience, and what are they looking for online?

What problems do your customers have that your website can help solve (e.g., providing online purchasing, scheduling services, etc.)?

What is the primary action you want visitors to take when they land on your website (e.g., making a purchase, filling out a contact form, subscribing to a newsletter)?

Let's break these questions down further to help you craft a clearer strategy.

Defining Your Business Model and Services

Start by examining your business model. What are your core services or products, and how will a website help you deliver those to your audience? Consider how a website can enhance the customer experience with your offerings. For example:

E-commerce Businesses: An e-commerce site needs to prioritize ease of navigation, search functionality, and a seamless checkout process. Your customers want to find products quickly and complete their purchase with minimal friction.

Service-Based Businesses: A website for a local service provider, like a hair salon or cleaning service, should focus on booking appointments, showcasing reviews, and providing easy ways for customers to get in touch.

Understanding the specific needs of your business model helps you determine what pages and features your website must have.

Identifying Your Target Audience

Your website's design, functionality, and content will differ based on who your target audience is. Building a website without understanding who you're trying to attract is like throwing darts in the dark. Different customer bases have different online behaviors and preferences, and your website should be optimized to meet their needs.

Demographics: Are your customers young professionals, retirees, or families? Knowing this helps you design your website accordingly. Younger audiences may prefer fast, visually appealing sites with minimal text, while older demographics might prefer detailed information and easy navigation.

Location: Are you targeting a local audience or a broader, national, or even global one? If your business caters to a specific geographic region, include local SEO strategies (we'll cover SEO in depth in later chapters) and highlight information like operating hours, location, and contact details prominently.

Customer Needs: What specific problems or pain points are your customers facing? For example, a financial adviser might attract clients looking for retirement planning, while a bakery attracts customers seeking a specific dietary offering, like gluten-free products. Your website needs to clearly communicate how you solve these problems.

Knowing your audience allows you to create tailored content and design that resonates with your visitors, increasing the likelihood of conversions.

Setting Your Website's Goals

Now that you've evaluated your business model and target audience, it's time to set concrete goals for your website. Your goals will influence the structure, content, and functionality of the site.

Here are some common goals small businesses might have for their websites:

Lead Generation: If your primary objective is to capture leads, you need to focus on building trust and creating effective calls-to-action (CTAs). Lead generation sites often feature contact forms, newsletter sign-ups, and valuable content such as blog posts or downloadable guides.

Online Sales: For businesses looking to sell products online, your primary goal should be creating an intuitive shopping experience. This includes ensuring a secure payment gateway, a smooth checkout process, and product descriptions that provide all necessary information to buyers.

Brand Awareness: Some businesses may simply want to build awareness and credibility within their industry. If your goal is to establish your expertise, you might focus on content marketing (like blogs, case studies, and white papers), and making sure your website communicates your brand's story effectively.

Clearly defining your website's purpose helps ensure that every element of the site is working toward that goal.

Competitor Research: Learning from Others

Competitor research is a vital part of the planning process. Studying your competitors' websites can provide valuable insights into what works in your industry—and what doesn't. By evaluating other businesses' online strategies, you can gain inspiration and avoid mistakes.

Here's how to conduct effective competitor research:

Identifying Competitors

Start by identifying who your direct competitors are. These are businesses offering the same or similar products/services in the same market. You can also look at businesses that may not be direct competitors but serve the same target audience. For example, if you own a bakery, you might look at how local restaurants with strong dessert offerings market themselves online.

Pro Tip: Use tools like Google Search and Yelp to find businesses in your industry and region. Make a list of 5-10 websites that appear to be performing well.

Analyzing Competitor Websites

Once you've identified your competitors, take a deep dive into their websites. Pay attention to the following elements:

Design and User Experience (UX): How is the website laid out? Is it easy to navigate, or is it cluttered and confusing? What features make the site user-friendly (e.g., clear menus, easy-to-read fonts, fast load times)?

Content: What kind of content do they provide? Are they focusing on product descriptions, blog posts, or educational resources? How do they showcase their services or products?

Call-to-Action (CTA): Are there clear CTAs encouraging visitors to book an appointment, request a quote, or buy a product? Are these CTAs prominent and persuasive?

SEO and Keywords: Look at the keywords your competitors are using. What terms appear frequently on their websites? You can use tools like Google's Keyword Planner or SEMrush to find which keywords are driving traffic to their sites.

Social Proof and Testimonials: Do they feature customer reviews or testimonials? How are they building trust with their audience?

Taking note of what works well on competitor websites and what falls short will help you improve your own site. It can also reveal gaps in the market that you can capitalize on, allowing you to stand out.

Matching Your Business Needs to Website Features

With a clear understanding of your business, audience, and competitors, it's time to consider what features your website should have to meet your goals.

Essential Features for Lead Generation

Contact Forms: Simple and easy-to-use forms that allow visitors to get in touch with you.

CTA Buttons: Prominent buttons that lead visitors to key actions like scheduling a call, downloading a brochure, or signing up for a newsletter.

Landing Pages: Dedicated pages optimized to capture leads, usually paired with an offer (e.g., free consultation, downloadable guide).

Must-Have Features for E-Commerce

Product Pages: Clear product descriptions with high-quality images, pricing details, and any necessary specifications.

Shopping Cart & Checkout: A secure, streamlined checkout process with multiple payment options (credit card, PayPal, etc.).

Customer Accounts: Offer users the option to create accounts for easier future purchases and order tracking.

Important Features for Service-Based Businesses

Appointment Scheduling Tools: Allow customers to book services online with real-time availability.

Testimonial Section: A dedicated area where you can showcase customer feedback, building trust.

Service Pages: Clear breakdowns of the services you offer, along with pricing and FAQs to address common customer concerns.

Once you have a list of features that match your goals, you can prioritize them based on your budget and technical capacity.

Website Types and Features

Choosing the Right Website for Your Business

When it comes to creating a website for your small business, one size definitely doesn't fit all. The type of website you choose will largely depend on the nature of your business, your goals, and how you want customers to interact with your site. Whether you need an online store to sell products, a service-oriented site that helps with appointment bookings, or a simple informational page, the features and layout of your website should reflect your specific needs.

This chapter will walk you through the various types of business websites and the must-have features that go along with each. By the end of this chapter, you'll have a clear understanding of the website type that's best suited to your goals.

Overview of Different Types of Business Websites

There are many types of websites out there, but small businesses generally fall into a few main categories. Understanding the different types will help you decide which format works best for your needs. Here's a breakdown of the most common website types for small businesses:

E-Commerce Websites

E-commerce websites are designed for businesses that sell products or services online. Whether you're a clothing boutique, a craft store, or even a restaurant offering take-out, an e-commerce site allows you to showcase products and process transactions directly through your website.

Key Features for E-Commerce Websites:

Product Catalog: A well-organized product catalog with images, descriptions, pricing, and inventory tracking is essential.

Shopping Cart: A secure shopping cart that allows customers to add multiple items before checking out.

Payment Gateway Integration: E-commerce sites need secure payment options such as credit cards, PayPal, or even Apple Pay.

Shipping and Tax Calculations: For product-based businesses, automatic shipping and tax calculations make the checkout process smoother for customers.

Customer Accounts: Offer customers the ability to create accounts for easy reordering, tracking shipments, and viewing past purchases.

SEO and Mobile Optimization: It's important to ensure that product pages are optimized for search engines and that the site is mobile friendly, as many users will shop on their phones.

An e-commerce website needs to prioritize user experience and simplicity. The easier it is for customers to navigate your store and make purchases, the better your conversion rates will be.

Service-Based Websites

For businesses that provide services—such as consultants, salons, or fitness trainers—your website's primary goal is to connect customers with the services you offer. These websites focus on showcasing your expertise, explaining your services, and making it easy for customers to book appointments or contact you.

Key Features for Service-Based Websites:

Service Descriptions: Provide clear, detailed explanations of each service you offer, including pricing and what customers can expect.

Appointment Scheduling: Online booking systems allow customers to check availability and schedule appointments at their convenience.

Client Testimonials: Displaying reviews from satisfied customers builds trust and credibility.

Contact Forms: Make it easy for potential customers to reach out with inquiries or requests for quotes.

Call-to-Action (CTA) Buttons: Encourage visitors to take the next step, whether it's scheduling a consultation, calling your office, or sending an inquiry.

Service-based websites benefit from clear navigation, prominent calls-to-action, and an intuitive user experience. The goal is to guide users from browsing to booking.

Portfolio Websites

Portfolio websites are commonly used by freelancers, creative professionals, or service providers who need to showcase examples of their work. For photographers, designers, architects, or even marketing professionals, these websites serve as an online gallery that helps attract new clients. Key Features for Portfolio Websites:

Visual Gallery: High-quality images or videos of your previous work organized in a way that highlights your best projects.

Case Studies or Descriptions: For more complex services, case studies or detailed descriptions of your work help potential clients understand your approach.

Testimonials: Client feedback adds credibility to your portfolio and helps build trust with prospective clients.

Contact Information: Make it easy for people to get in touch to inquire about hiring you.

A clean, visually engaging design is crucial for portfolio websites since the main goal is to showcase your work in the best light possible.

Informational Websites

Many small businesses, particularly local businesses like restaurants, law firms, or dental practices, may not need e-commerce functionality or elaborate booking systems. Instead, their primary goal is to provide information—things like operating hours, services offered, or contact information. These websites serve as digital brochures that give visitors the key information they need to engage with your business.

Key Features for Informational Websites:

Business Overview: Provide a clear summary of what your business does and what makes it unique.

Contact Details: Ensure your phone number, email address, and physical address are easy to find.

Service or Product Information: Include a section that describes the services you offer or the products you sell.

FAQs: A section answering common customer questions can help reduce phone calls and inquiries.

Blog or News Section: Some businesses include a blog to provide updates or share news about the business or industry. This can also help with SEO.

Informational websites are simple yet highly functional, focusing on helping customers find the basic information they need to make decisions about engaging with your business.

Blogs or Content-Based Websites

Some businesses thrive by providing content. Whether it's a blog, an educational resource, or an industry news site, content-based websites prioritize delivering valuable information. Businesses in industries like media, marketing, education, or coaching often rely on a blog or content-heavy website to build an audience and create a following.

Key Features for Blogs or Content-Based Websites:

Blog Posts or Articles: A regularly updated section for new content, categorized and easy to browse.

Content Sharing Options: Integration with social media or email sharing options to help readers distribute your content.

Newsletter Sign-Ups: Capture visitor emails with subscription forms, so you can build an audience and stay in touch.

Search Functionality: A robust search bar to help users find specific content quickly.

Related Content Suggestions: Display related posts or articles to encourage further browsing.

Content-based websites are powerful tools for building authority in your industry and nurturing relationships with your audience.

Deciding Which Type Is Right for You

Now that you're familiar with the different types of business websites, it's time to decide which is best for your small business. Here are a few guiding questions to help you make your choicc:

What Is the Main Function of Your Website?

If your goal is to sell products, an e-commerce site is the obvious choice.

If you want to showcase your work to attract clients, a portfolio website is the way to go.

If you're focused on driving bookings for a service, a service-based website with online scheduling is key.

What Experience Do You Want to Provide to Visitors?

Do you want a simple, informational site that tells people who you are and how to contact you?

Do you need to create an immersive experience with visual galleries and multimedia?

Would you like to build a resource for your audience with articles, blog posts, and other educational materials?

Resources for Website Maintenance

Some websites, such as e-commerce and content-heavy sites, require regular updates and maintenance, while informational sites may be more static and require less frequent changes. Consider your ability to keep your site up-to-date, both in terms of content and technical upkeep.

Custom vs. Template Websites: Pros and Cons

Another decision you'll need to make is whether to build a custom website from scratch or use a template. Each option has its pros and cons depending on your budget, timeline, and specific needs.

Custom Websites

Custom-built websites are designed and coded specifically for your business. This option offers full control over every aspect of the site, from design to functionality.

Pros:

Tailored to Your Business: You can ensure the site matches your branding and provides the exact features you need.

Scalable: Custom sites are easier to scale as your business grows.

Unique Design: A custom website ensures your site doesn't look like anyone else's site.

Cons:

Cost: Custom websites are typically more expensive to build due to the development and design time involved.

Time-Consuming: The development process can take longer compared to using a template.

Ongoing Maintenance: Custom sites may require a developer for ongoing updates and fixes.

Template Websites

Template websites, also called "theme-based" websites, use pre-designed layouts and structures that can be customized to a certain extent.

Pros:

Cost-Effective: Templates are usually much cheaper than building a site from scratch.

Faster to Launch: You can get a website up and running quickly, sometimes in just a few days.

Easy to Use: Many template platforms are designed for nontechnical users, allowing you to make changes without needing a developer.

Cons:

Limited Customization: You may be limited in terms of layout, design, and functionality.

Less Unique: Since many businesses may use the same template, your site may look similar to others.

Scalability: Some template sites can be difficult to scale as your business grows and your needs evolve.

Your Website, Your Way

Choosing the right type of website is a crucial step in building your business's online presence. Whether you're selling products, showcasing services, or providing information, the right website will help you connect with your target audience and achieve your business goals. With the right features, design, and a trusted partners you can create a website that not only looks great but performs effectively for your business.

Budgeting for Your Website

Why Budgeting Matters for Your Website

Building a website is one of the most valuable investments a small business owner can make, but it's an investment that requires careful planning and budgeting. Whether you're creating a simple informational site or a fully integrated e-commerce platform, the costs associated with building, maintaining, and optimizing your website can vary widely.

Many small business owners underestimate the total cost of owning a website, focusing only on the initial setup while overlooking ongoing expenses like hosting, updates, and marketing. In this chapter, we'll break down the various elements of website budgeting, show you where you can save money, and help you avoid common pitfalls that lead to overspending.

By the end of this chapter, you'll have a clear understanding of how to budget effectively for your website, ensuring you get the best possible return on your investment without sacrificing quality.

Key Cost Components of a Website

The total cost of a website can be divided into several key components. It's important to understand what each of these components involves so that you can plan your budget effectively.
Let's break them down:

Domain Registration

Your domain name is your website's address (e.g., www. yourbusiness. com), and it's often the first thing potential customers will see. A memorable and professional domain name can make a big difference in your online presence, so choosing the right one is crucial.

Costs Involved:

Initial Purchase: Domain names typically cost between $10 - $50 per year or more, depending on the extension (.com, .net, .org, etc.) and the popularity of the name.

Renewal Fees: Domain registration is an ongoing cost, and renewal fees are often required yearly.

Premium Domains: If the domain you want is highly sought after or already owned, you may need to purchase it from a third party, which can cost hundreds or even thousands of dollars.

Tips for Saving Money:

Choose a domain name that is simple but unique enough to avoid high costs.

Domain registration services are available at competitive rates, making it easy to secure a great domain for your business.

Web Hosting

Hosting is what keeps your website online and accessible to visitors. Think of it as the digital space where your website lives. Hosting services vary widely in terms of cost and features, and the type of hosting you choose can significantly impact your website's performance. Types of Hosting:

Shared Hosting: This is the most affordable option, where your website shares server resources with other websites. Suitable for smaller websites with moderate traffic.

VPS (Virtual Private Server) Hosting: A middle-ground option that offers more resources and flexibility than shared hosting, but at a higher cost. Ideal for businesses expecting moderate to high traffic.

Dedicated Hosting: The most expensive option, where your website has an entire server to itself. This is best for large businesses with heavy traffic or complex site needs.

Costs Involved:

Shared hosting typically costs between $5 and $15 per month.

VPS hosting ranges from $20 to $100+ per month, depending on the level of resources you need.

Dedicated hosting can range from $100 to $500+ per month, depending on server specifications.

Tips for Saving Money:

Start with shared hosting if your website is small, and upgrade as your business grows.

Many hosting providers offer affordable and scalable hosting plans that grow with your business, ensuring you don't overpay for services you don't need.

Website Design and Development

Design and development are where the bulk of your upfront costs will likely occur. The look, feel, and functionality of your website all depend on this stage, and it's important to allocate a significant portion of your budget here, especially if you want a custom-built site.

Options for Design and Development:

DIY Website Builders: Tools like Wix, Squarespace, and WordPress offer easy-to-use platforms for building websites without needing coding skills. While affordable, they may lack the customization options and scalability that businesses eventually need.

Hiring a Freelance Designer/Developer: This is a good middle ground for businesses that want more control over the design but don't need a large-scale website. Costs can vary widely depending on the experience of the freelancer.

Hiring a Professional Agency: For businesses that require a fully custom website or complex features, hiring an agency is the best route, though it can be expensive.

Costs Involved:

DIY website builders typically cost $10 to $50 per month, depending on the platform and features.

Freelance designers may charge anywhere from $500 to $5,000, depending on the scope of the project.

Agencies often start around $5,000 and can go up to $20,000+ for larger, more complex projects.

Tips for Saving Money:

Start small with a DIY platform if you're on a tight budget, then consider upgrading to professional services as your business grows.

Many hosts provide both DIY website options and professional design services, allowing you to start affordably and scale when needed.

Plugins, Features, and Custom Functionality

Your website may require additional features beyond the basic design. These can include e-commerce functionality, booking systems, forms, or other custom elements that enhance the user experience and drive conversions.

Costs Involved:

Free Plugins: Many platforms like WordPress offer free plugins for basic features.

Premium Plugins: More advanced plugins (such as e-commerce solutions or membership tools) typically range from $20 to $200 per year.

Custom Development: If you need unique functionality that can't be covered by existing plugins, custom development costs can vary widely depending on complexity. Expect to pay anywhere from $500 to several thousand dollars.

Tips for Saving Money:

Look for free or low-cost plugins that meet your needs, and only invest in custom development when absolutely necessary.

Web hosting companies offer expert advice on selecting the right plugins and features to enhance your site without overspending.

Ongoing Maintenance and Updates

Your website isn't a one-time project. It requires regular updates, maintenance, and optimization to ensure it continues to perform well and remain secure. This is an often-overlooked cost that can catch business owners off guard.

Costs Involved:

Security Updates and Patches: Essential for protecting your website from hackers and ensuring it stays compliant with the latest web standards.

Content Updates: Adding new content or refreshing old content to keep your site relevant.

Website Optimization: Ensuring that your website loads quickly and performs well as traffic grows.

Tips for Saving Money:

Consider a monthly maintenance package from a fully managed web hosting provider. These types of companies offer website maintenance services, including security updates and performance optimization, for a predictable, budget-friendly cost.

DIY vs. Hiring a Professional: Cost Comparison

Many small business owners face a tough decision: should you build your website yourself using a DIY platform or hire a professional to do it for you? The answer depends largely on your budget, time constraints, and the complexity of the website you need.

DIY Website Builders:

Pros: Lower upfront cost, quick to launch, easy for non-technical users.

Cons: Limited customization, potential for "cookie-cutter" design, often requires upgrades for advanced features.

If you're starting small and need a basic website, DIY builders like Wix or Squarespace are great options. These platforms typically offer tiered pricing, so you can start with a low-cost plan and upgrade as your needs grow. However, if you need a highly customized website, DIY builders may fall short of delivering the flexibility you require.

Hiring a Professional:

Pros: Custom design and functionality, scalable, better for SEO and long-term growth.

Cons: Higher upfront costs, longer development timeline, ongoing expenses for updates and changes.

Hiring a professional (either a freelance developer or a design agency) is the best option for businesses that need a more sophisticated website or have specific branding and functionality requirements. While the upfront cost is higher, a custom-built website is more likely to meet your exact business needs and is easier to scale over time. Tips for Saving Money:

Start with a simple DIY solution and migrate to a professional service when your business can support the cost.

Many hosts offer hybrid solutions, allowing you to start with a DIY platform while also providing options for professional upgrades as your business grows.

How Much Should You Spend?

So how much should you actually spend on a website? The answer depends on your business size, the complexity of your site, and your long-term goals. However, here are some rough guidelines to help you plan:

Small, Local Businesses: A budget of $500 to $2,000 is often enough to get a simple, functional website off the ground. This could include using a DIY platform or working with a freelancer.

Medium-Sized Businesses: If you need more advanced features like e-commerce, online booking, or custom functionality, plan to spend $2,000 to $10,000. This could include working with a professional developer or design agency.

Large Businesses or Complex Sites: For businesses that need a custom-built website with many features, plan on budgeting $10,000 to $50,000 or more. This includes not only the initial design and development but also the long-term maintenance and scalability of the site.

Pro Tip: Don't forget to account for ongoing costs like hosting, maintenance, and marketing (SEO, advertising, etc.) in your budget. Websites are long-term investments, and ongoing expenses are a part of keeping your site effective.

Getting the Best Value for Your Money

No matter your budget, there are always ways to get the most value for your money. Here are some practical tips for maximizing your website investment:

Prioritize Must-Have Features

It's easy to get caught up in all the cool features and add-ons, but focus on what's essential for your website's success. If you're on a tight budget, prioritize the basics first: good design, clear navigation, mobile optimization, and fast load times. You can always add more features down the road.

Negotiate with Professionals

If you're working with a freelance designer or agency, don't be afraid to negotiate pricing or ask for a payment plan. Many professionals offer flexible payment structures that can help you manage your cash flow while still getting a high-quality website.

Plan Your Budget, Maximize Your Investment

Budgeting for your website is a critical step in ensuring your business's online success. By understanding the key cost components, comparing DIY options with professional services, and planning for ongoing expenses, you can create a realistic budget that helps you build a website that meets your goals without breaking the bank.

With a clear budget and the right strategy, your website can become one of the most valuable assets for your business, helping you attract customers, drive sales, and grow your brand. You can rest assured that your investment is well spent, and your website will be built to last.

Choosing the Right Domain and Hosting

The Foundation of Your Website

Just as a physical business needs a location and a sign that makes it easy for customers to find, a website needs two foundational elements: a domain name and hosting. Without these, a website wouldn't exist online. The domain name is your website's address—the way people find you—while the hosting service is the infrastructure that keeps your website live and accessible.

Choosing the right domain name and hosting service can have a major impact on your website's success. A good domain name helps build your brand and makes it easy for customers to find you, while the right hosting plan ensures your site runs smoothly and securely.

In this chapter, we'll break down the essentials of choosing a domain name, explain the different types of hosting, and provide tips on how to make the best choices for your business.

Choosing the Perfect Domain Name

Your domain name is your website's address on the internet, and it's one of the most important parts of building your online presence. It needs to be easy to remember, relevant to your business, and help with search engine optimization (SEO). Whether it's a simple name or something creative, your domain name will serve as a key piece of your branding strategy.

What Makes a Good Domain Name?

A good domain name should be simple, memorable, and reflective of your business. Here are some key guidelines to follow when selecting the perfect domain name:

Keep It Short and Simple: A domain name that's easy to type and spell will make it easier for customers to find you. For example, if you own a landscaping business, something like "GreenGarden. com" is much easier to remember than "TheBestGardenLandscapingCompanyInTheArea.com."

Avoid Numbers and Hyphens: Numbers and hyphens often lead to confusion. People might not remember whether they need to spell out numbers or use digits, and hyphens are often forgotten, leading to potential traffic loss.

Use Keywords: Including relevant keywords in your domain name can help with SEO. If your business is local or industry-specific, try including keywords related to your location or services. For example, a bakery in Miami could benefit from a domain like "MiamiCakes.
com."

Make It Brandable: Your domain name is a reflection of your brand. Ideally, it should be unique enough to stand out and easy to associate with your business. Many successful businesses choose to use their company name as the domain (e.g., "JoesPlumbing.com"), which helps strengthen brand recognition.

Choose the Right Extension (.com, .net, .org): The .com extension is the most common and trusted, but there are many other options like .net, .org, and .co. If possible, always go for .com, as it's the most familiar to users, but don't shy away from alternatives if your preferred domain is taken.

Domain Availability and Alternatives

It's common to find that your ideal domain name is already taken. In that case, don't be discouraged—there are several ways to find a great alternative:

Try Variations of Your Name: If "GreenGarden.com" is taken, try "GreenGardenServices.com" or "GreenGardenPros.com." Adding simple words like "services" or "solutions" can make your domain unique while still being relevant to your business.

Use a Different Extension: If the .com version of your domain is unavailable, consider using .net, .biz, or a local extension like .us. This is particularly useful for regional businesses looking to capture a local market.

Purchase a Premium Domain: Sometimes, the perfect domain is already owned but available for purchase. These "premium domains" are often sold by domain registries or private owners. While premium domains can be expensive, they are sometimes worth the investment if they closely match your brand and are highly memorable.

Check Domain History: Before finalizing your domain, it's important to ensure that it doesn't have a negative history (e.g., it wasn't previously associated with spam or malicious activity). Tools like Wayback Machine or Whois can help you check a domain's past.

Domain Registration: Getting It Done Right

Once you've decided on a domain name, the next step is registering it. Registering a domain means securing the rights to use that domain name for your website. You can do this through domain registration services, often bundled with hosting providers.

Steps to Register a Domain:

Search for Availability: Use domain registration platforms like GoDaddy or Namecheap to check if your desired domain is available.

Register Your Domain: Once you find an available domain, you'll register it under your name or your business's name. This gives you ownership of the domain for a specified period (typically one year, but you can often register for multiple years).

Renew Regularly: Make sure to renew your domain registration before it expires to avoid losing it. Most registration platforms offer automatic renewal to prevent lapses.

Pro Tip: Most hosting companies offer domain registration services, helping small businesses secure the perfect domain with ease and at competitive prices.

Understanding Website Hosting

If the domain is your website's address, hosting is the actual space where your website's files and data are stored. It's the digital "real estate" that powers your website and makes it available to visitors. Choosing the right hosting plan is essential to ensure your website runs smoothly, loads quickly, and stays secure.

What Is Website Hosting?

Web hosting involves storing your website's files—such as text, images, and videos—on a server that is connected to the internet. When someone types in your domain name, the hosting service retrieves these files and displays your website to the visitor.

Hosting plans come in different shapes and sizes, with varying levels of speed, security, and scalability. The type of hosting you choose will depend on your website's size, traffic expectations, and technical needs.

Types of Website Hosting

There are several types of hosting, each suited for different kinds of businesses. Here's a quick overview of the main options:

Shared Hosting

What It Is: In shared hosting, your website shares a server with other websites. This is the most affordable option and is suitable for smaller websites with lower traffic.

Pros: Low cost, easy to set up, great for small businesses just starting out.

Cons: Limited resources (since you share the server with others), slower speeds, and less flexibility for scaling.

VPS (Virtual Private Server) Hosting

What It Is: VPS hosting offers a dedicated portion of a server's resources, meaning you get more power and flexibility compared to shared hosting. It's a good option for businesses expecting moderate traffic or needing more control over server settings.

Pros: More resources and control than shared hosting, better performance.

Cons: Higher cost than shared hosting, more technical knowledge may be needed for setup and maintenance.

Dedicated Hosting

What It Is: Dedicated hosting provides an entire server solely for your website. It's ideal for businesses with high traffic or complex websites that require significant resources.

Pros: Full control over the server, highest level of performance and security.

Cons: Expensive, requires technical expertise to manage.

Cloud Hosting

What It Is: Cloud hosting uses multiple servers to host your website, allowing for greater flexibility and scalability. Your site isn't tied to one physical server, so it can handle traffic spikes more easily.

Pros: Scalability, reliability (if one server fails, others pick up the slack), good for growing businesses.

Cons: Can be more expensive than other options, complexity may require professional management.

Managed WordPress Hosting

What It Is: If you're using WordPress to build your site, managed WordPress hosting takes care of many technical aspects such as updates, security, and backups. It's a popular choice for small businesses that want a WordPress site without dealing with technical maintenance.

Pros: Optimized for WordPress, hassle-free maintenance.

Cons: Generally limited to WordPress sites, can be more expensive than shared hosting.

Key Factors to Consider When Choosing a Hosting Provider

When selecting a hosting provider, there are a few important factors to keep in mind:

Uptime: Uptime refers to the percentage of time your website is accessible online. Most hosting providers offer a 99.9% uptime guarantee, but even a slight difference can impact your site's availability to visitors.

Speed: A slow-loading website can hurt user experience and SEO rankings. Ensure your hosting provider offers fast load times, as this is critical for customer satisfaction and search engine visibility.

Security: Your hosting provider should offer robust security features, including firewalls, SSL certificates, and regular backups to protect your site from hackers.

Scalability: If you expect your website to grow in the future, choose a hosting plan that allows for easy upgrades and scalability.

Pro Tip: Web hosting companies offer a range of hosting plans tailored to small business needs, with reliable uptime, fast load speeds, and top tier security features.

Why Hosting and Domain Choices Impact SEO and User Experience

While it might seem like domain names and hosting are simply technical necessities, they actually play a significant role in your website's SEO and user experience.

SEO (Search Engine Optimization)

Your domain name can impact your SEO performance. Search engines like Google take into account domain names and hosting when determining where to rank websites. Here's how:

Keywords in Domain Names: If your domain name includes relevant keywords, it can help boost your search engine rankings. For example, if you're a plumber in Chicago, a domain like "ChicagoPlumbingServices.com" can give you a slight edge in SEO.

SSL Certificates: Having an SSL certificate (which secures your website and gives you the "https://" instead of "http://") is now a ranking factor for Google. Most hosting providers, include SSL certificates in their packages.

Server Location and Speed: Google values fast-loading websites. Hosting providers with servers located closer to your audience can reduce loading times and improve your rankings.

User Experience

Loading Speed: Visitors expect websites to load quickly—typically within two to three seconds. If your hosting provider can't deliver fast loading times, you'll likely lose visitors before they even see your content. In fact, studies show that 40% of people will leave a website if it takes longer than three seconds to load.

Downtime: Frequent website downtime can frustrate users and hurt your credibility. If your site is often inaccessible due to poor hosting, you could lose business and damage your reputation.

Mobile Optimization: With the increasing use of mobile devices to browse the web, it's essential that your website works smoothly on smartphones and tablets. Many hosting providers, offer mobile optimization tools to ensure your site looks and performs well across all devices.

Building a Solid Foundation for Your Website

Choosing the right domain name and hosting provider is essential for building a website that not only looks good but performs well. These two elements form the foundation of your online presence, influencing everything from SEO to user experience.

DIY vs. Hiring a Professional Web Designer

The Big Decision

One of the most important decisions you'll face when creating your business website is whether to build it yourself using a DIY website builder or hire a professional web designer (or agency) to handle it for you. Both options have their advantages and challenges, and the right choice will depend on your budget, technical skills, and long-term goals. In this chapter, we'll explore the key differences between DIY website builders and hiring a professional, breaking down the pros and cons of each. By the end, you'll have a clear understanding of which route is best for your business.

DIY Website Builders: An Affordable, Hands-On Approach

For small business owners on a tight budget or those who like to be hands-on, DIY website builders like Wix, Squarespace, and WordPress are popular options. These platforms offer a range of pre-designed templates and easy-to-use tools, allowing non-technical users to create a website without knowing how to code.

But is a DIY website builder the right fit for your business? Let's look at the benefits and potential drawbacks.

Pros of DIY Website Builders

Cost-Effective

One of the main reasons small businesses choose a DIY website builder is affordability. Most DIY platforms offer monthly or annual subscriptions that range from as low as $10 to $50 per month, depending on the features you need.

Basic Plans: DIY platforms usually offer free or low-cost plans that provide basic features like hosting, domain connection, and pre-designed templates.

Affordable Upgrades: As your business grows, you can upgrade to premium plans that offer additional functionality, such as ecommerce, SEO tools, and integrations with other apps.

Quick and Easy to Set Up

DIY builders are designed to be user-friendly, with drag-and-drop interfaces that allow you to build your site without any coding skills. This means you can have a website up and running within a matter of hours or days, rather than weeks or months.

Templates: Most platforms offer a wide range of pre-designed templates that are customizable, so you can choose a look that fits your brand.

Instant Changes: If you need to make a quick update—such as adding a new product or changing business hours—you can easily log in and make the change yourself.

Creative Control

Using a DIY builder gives you full creative control over your website's design and content. Since you're building it yourself, you can experiment with different layouts, colors, and features until you're happy with the final result.

No Long-Term Commitments

Most DIY website builders operate on a subscription basis, which means you're not locked into a long-term contract. This flexibility is ideal for businesses that want to keep costs low in the early stages and upgrade their site as they grow.

Cons of DIY Website Builders

Limited Customization

While DIY builders offer flexibility, there are limitations. You're restricted by the templates and features provided by the platform, which can lead to your site looking similar to others that use the same service.

Template Limitations: Many DIY templates come with predefined layouts and sections, meaning you might not have complete control over certain design elements.

Functionality Constraints: If your business needs specific features like complex booking systems, custom e-commerce functionalities, or tailored integrations, DIY platforms may fall short.

Time-Consuming

Building a website, even with a user-friendly tool, can still be time consuming—especially if you're learning as you go. Business owners often underestimate the time it takes to design, write content, and maintain a website.

Learning Curve: Even though DIY builders are easy to use, there's still a learning curve when it comes to designing an effective site, optimizing for SEO, and managing content.

SEO and Performance Limitations

DIY platforms often come with basic SEO tools, but they may lack the advanced features necessary to optimize your site for search engines effectively. Additionally, the performance (speed, mobile responsiveness) of DIY sites may not match those built by professionals, which can affect user experience and SEO rankings.

Template Optimization: Not all templates are created equal. Some might look visually appealing but could slow down your site if they aren't optimized for performance.

Limited Control Over SEO: DIY platforms typically handle SEO basics, like meta tags and alt text, but they don't always allow for advanced SEO customizations, such as structured data or comprehensive analytics tracking.

Scalability Issues

As your business grows, you might find that your DIY site struggles to keep up with your needs. Expanding your site's functionality or performance may require switching to a more advanced platform or hiring a professional to migrate your site, which can add unexpected costs.

Hiring a Professional Web Designer: A Tailored, Hands-Off Approach

On the other end of the spectrum, hiring a professional web designer or design agency provides a more customized, hands-off approach. A professional will work with you to create a website that's tailored to your business's unique needs, offering more flexibility and control over the final product.

Let's dive into the advantages and disadvantages of this approach.

Pros of Hiring a Professional Web Designer

Customization and Flexibility

One of the biggest advantages of working with a professional is that you'll get a website that is fully customized to your brand and business goals. A professional designer can create a unique look and feel that aligns with your company's identity and helps you stand out from competitors.

Custom Design: You're not limited to pre-made templates. Every aspect of your website can be tailored to your vision, whether that means custom graphics, interactive features, or a completely unique layout.

Functionality: Professional designers can incorporate advanced features that are difficult to achieve with DIY builders, such as complex e-commerce platforms, member portals, or integrations with specific software tools your business uses.

Better Performance and SEO

When you hire a professional, you're also investing in the performance of your site. Professionals understand how to build websites that load quickly, look great on mobile devices, and are optimized for search engines. This can lead to better user experiences and higher rankings in search results.

SEO Best Practices: A professional web designer or agency will often incorporate advanced SEO strategies from the ground up, such as proper site architecture, internal linking, and optimized page speeds.

Mobile Optimization: Ensuring that your site looks and functions well on mobile devices is critical, and professionals will build mobile responsive sites that adapt seamlessly across different screen sizes.

Time-Saving

Designing a website from scratch takes time. By hiring a professional, you free up valuable time to focus on running your business while leaving the technical details to an expert. From initial design to ongoing maintenance, a professional can handle the entire process, ensuring your site is always up to date and functional.

Support and Maintenance

Many professional designers and agencies offer ongoing support and maintenance packages. This means you'll have someone to turn to if something breaks, if your site needs updates, or if you want to add new features in the future.

Regular Updates: Websites need to be regularly updated for security, functionality, and content. A professional can handle these updates for you.

Troubleshooting and Support: If you run into technical issues, a professional designer can resolve them quickly and efficiently, preventing downtime and loss of business.

Cons of Hiring a Professional Web Designer

Higher Upfront Costs

The most significant downside to hiring a professional web designer or agency is the higher cost. Custom-built websites can range from a few thousand dollars to $10,000 or more, depending on the complexity of your needs.

Initial Investment: You're paying for expertise, custom design, and advanced functionality, which can be expensive for small businesses with limited budgets.

Ongoing Costs: In addition to the upfront cost, you may need to pay for ongoing support, hosting, and maintenance services.

Longer Development Timeline

Building a website from scratch takes time. Depending on the scope of the project, it can take weeks or even months to go through the design, development, testing, and launch phases. If you need a site up quickly, the timeline might not align with your needs.

Less Immediate Control

With a professional designer handling the project, you may have less direct control over the day-to-day building process. While you'll be involved in major decisions, the technical aspects will be handled by the designer, which means you'll need to communicate clearly to ensure the final result aligns with your vision.

How to Decide: DIY or Professional?

Deciding whether to go the DIY route or hire a professional web designer depends on several factors, including your budget, business needs, and long-term goals. Let's break down the decision-making process:

When to Choose a DIY Website Builder A DIY website builder is a good option if:

You Have a Limited Budget: If your priority is getting a website up quickly and affordably, a DIY platform is a great choice.

You Need a Simple Website: If your website is primarily informational (e.g., a brochure site), and doesn't require complex functionality like e-commerce or custom integrations, a DIY builder will suffice.

You Want Full Creative Control: If you like the idea of building your site yourself and having complete control over every aspect, a DIY platform will give you the tools to experiment.

You Don't Mind Regular Maintenance: If you're comfortable with making updates and changes on your own, you'll save money by handling maintenance yourself.

When to Hire a Professional Web Designer

Hiring a professional is the best option if:

You Need a Complex Website: If you need a site with advanced functionality, such as custom e-commerce features, a professional can create a tailored solution that meets your exact needs.

You Want a Unique, Custom Design: If branding and aesthetics are critical to your business, a professional can create a website that stands out and is fully aligned with your vision.

You Want to Save Time: If you don't have the time or expertise to build and maintain a website yourself, hiring a professional will allow you to focus on running your business.

You Have Long-Term Growth in Mind: If you expect your website to grow or evolve over time, a professional designer can build a site that is scalable and adaptable to your changing needs.

Making the Right Choice for Your Business

Choosing between a DIY website builder and hiring a professional web designer is one of the most important decisions you'll make when building your online presence. Both options have their benefits, and the right choice will depend on your business's unique needs, budget, and long-term goals.

By considering the pros and cons of each approach, and with the help of your web development and hosting company, you can create a website that meets your needs today and grows with your business in the future.

Planning and Designing the Website

Why Planning Is Key

Building a website without a plan is like trying to build a house without a blueprint. A well-structured, thoughtful approach to website planning and design ensures that your site not only looks good but functions in a way that meets your business goals. Whether you're building it yourself or working with a professional designer, having a clear plan for your website's structure, design, and content is essential.

In this chapter, we'll explore how to effectively plan your website's structure, create a user-friendly design, and make strategic decisions about branding and layout. By the end of this chapter, you'll have a road map for building a website that's both functional and visually appealing.

Defining Your Website's Structure: The Blueprint for Success

The first step in designing a website is to define its structure. This is also known as the site map—a visual or written representation of how your website's pages will be organized. A well-organized structure ensures that visitors can easily navigate your site and find the information they need.

Common Website Pages

While every website is unique, most business websites follow a similar structure. Here are the common pages that small business owners should consider including:

Homepage

Purpose: Your homepage is the first thing most visitors will see. It needs to make a strong impression and guide visitors to the information they're looking for.

Key Elements: A clear headline, brief explanation of your business, navigation to key sections, and a call to action (e.g., "Contact Us" or "Learn More").

About Us

Purpose: The "About Us" page tells visitors who you are, what you do, and why they should trust you. It's an opportunity to share your brand's story and build credibility.

Key Elements: A brief history of your business, core values, key team members, and any certifications or achievements.

Services or Products

Purpose: This page details what your business offers. If you sell products, this is where your product listings will be. If you offer services, you'll describe each service in detail.

Key Elements: Clear descriptions of services or products, pricing (if applicable), and any additional information like delivery or booking options.

Contact Us

Purpose: The "Contact Us" page provides all the ways customers can reach you.

Key Elements: Include a contact form, your business address, phone number, email, and links to social media channels.

Blog or News

Purpose: If you plan to engage in content marketing, your blog or news section will feature updates, articles, or industry insights that help attract and retain customers.

Key Elements: Regularly updated content, easy navigation between posts, and social sharing buttons.

FAQs

Purpose: The FAQs (Frequently Asked Questions) page addresses common questions customers may have about your products, services, or policies. This reduces the number of inquiries you need to respond to manually.

Key Elements: Simple, concise answers to common questions.

Prioritizing Pages Based on Your Business Needs

Not all businesses need every page. When deciding which pages are essential for your site, consider your specific goals. For example:

E-commerce businesses will need dedicated product pages, a shopping cart, and checkout functionality.

Service-based businesses may prioritize appointment booking features and detailed service descriptions.

Content-focused businesses might focus on a blog or portfolio section to showcase their expertise.

Once you've determined which pages are most important for your business, map them out in a simple hierarchy, starting with the homepage and working your way down. This site map will act as a guide for your design process.

User Experience (UX) Design: How to Make Your Website Easy to Use

Design isn't just about making your website look good—it's about creating a site that's easy to navigate and user-friendly. This is where User Experience (UX) Design comes into play.

A well-designed website should guide visitors effortlessly from one page to another, encouraging them to take action (e.g., make a purchase, book an appointment, or contact you).

Navigation: The Backbone of User Experience

Your website's navigation should be intuitive and straightforward. Visitors should be able to find what they're looking for in just a few clicks. Here's how to create an effective navigation system:

Keep It Simple: Limit your main navigation menu to the most important pages. Typically, the homepage, About Us, Services/ Products, and Contact pages should be in the main menu.

Use Clear Labels: Avoid using jargon or vague labels in your menu. For example, instead of "Solutions" or "Capabilities," use more straightforward terms like "Services" or "What We Do."

Consider Drop-Down Menus: If you have multiple services or product categories, use drop-down menus to organize sub-pages. However, keep the menu simple and avoid overwhelming visitors with too many options.

Call to Action (CTA): Guiding Visitors to Take Action

Every page on your website should include a Call to Action (CTA) that guides visitors toward the next step, whether it's making a purchase, filling out a contact form, or signing up for a newsletter. CTAs should be clear, concise, and visually prominent.

Best Practices for CTAs:

Be Direct: Use action-oriented language like "Buy Now," "Get a Free Quote," or "Contact Us."

Place CTAs Strategically: Make sure your CTA buttons are placed where users naturally look—such as the top right corner of the navigation bar, at the end of a service description, or within a prominent section on the homepage.

Use Contrast: Make your CTA buttons stand out by using contrasting colors. They should be easy to spot at a glance.

Mobile Responsiveness: Designing for All Devices

With the majority of web traffic now coming from mobile devices, it's critical that your website looks and functions just as well on smartphones and tablets as it does on desktop computers. This is known as mobile responsiveness, and it's essential for both user experience and search engine rankings.

Key Considerations for Mobile-Friendly Design:

Responsive Layout: Ensure your design automatically adjusts to different screen sizes. DIY platforms like Wix and Squarespace offer responsive templates, while professionals will build this functionality into custom designs.

Simplify Navigation: Mobile users should be able to navigate your site with their thumbs. Use a simple hamburger menu (the three-line icon) and ensure that buttons and links are large enough to be tapped easily.

Fast Load Times: Mobile users expect quick load times. Compress images and minimize the use of heavy features like videos or animations to improve mobile performance.

Creating a Consistent and Professional Look

Your website is often the first interaction potential customers will have with your brand, so it's essential that it reflects your business's identity. Consistent branding creates trust and makes your website more memorable.

Choosing Colors and Fonts

Your color palette and font choices should align with your brand's overall identity. For example, a law firm might opt for neutral, professional colors like navy and gray, while a children's toy store might use bright, playful colors like yellow and orange.

Tips for Choosing Colors:

Stick to Your Brand Colors: If you already have established brand colors (from your logo or marketing materials), use those on your website to maintain consistency.

Limit Your Palette: Choose 2-3 main colors and use them consistently across your site. Too many colors can make your site look cluttered and unprofessional.

Think About Contrast: Ensure there's enough contrast between text and background colors so that your site is easy to read. For example, use dark text on a light background or vice versa.

Tips for Choosing Fonts:

Keep It Simple: Choose no more than two fonts—one for headings and one for body text. Avoid overly decorative or hard-to-read fonts.

Ensure Readability: Make sure your font size is large enough to read easily on all devices. A minimum of 16px is recommended for body text.

Using High-Quality Images and Videos

Images and videos play a significant role in website design. They can help tell your brand's story, showcase your products, and create an emotional connection with your audience. However, poor-quality visuals can make your site look unprofessional, so it's important to use high-quality images and videos.

Best Practices for Using Visuals:

Use Original Photos: If possible, use original photos of your business, team, or products. Stock images are an option, but original content is more authentic and engaging.

Optimize for Web: High-resolution images can slow down your website, especially on mobile devices. Compress images before uploading them to ensure they load quickly.

Incorporate Video: Video content can boost engagement. Use videos to showcase your products, explain your services, or share customer testimonials.

Consistent Branding Across Platforms

Your website should reflect your brand in every aspect—from your logo and color scheme to your tone of voice and messaging. To ensure consistency, align your website's design with other marketing materials, such as social media profiles, business cards, and brochures.

Building a Website That Works for Your Business

Planning and designing a website is a critical part of establishing your online presence. By starting with a clear structure, focusing on user experience, and ensuring consistent branding, you can build a site that not only looks great but also helps achieve your business goals.

With today's design tools and services, you can create a website that reflects your brand, engages visitors, and converts them into customers. By following the strategies outlined in this chapter, you'll be well on your way to launching a successful website.

Creation: What Should Go On Your Website?

The Role of Content in Your Website's Success

Your website's content is the backbone of your online presence. It's what engages visitors, tells your brand's story, and persuades potential customers to take action. In fact, no matter how beautifully designed your website is, without great content, it won't fulfill its potential. Whether it's written copy, images, or video, your content needs to speak to your audience, reflect your brand, and provide valuable information.

In this chapter, we'll cover what kinds of content are essential for your website, how to create high-quality content that aligns with your business goals, and how to optimize your content for SEO. By the end, you'll have a clear understanding of what should go on your website to captivate visitors and keep them coming back.

Writing Effective Website Copy

The written content on your website—known as copy—plays a major role in how visitors perceive your brand and whether they choose to engage with your business. Whether it's the headline on your homepage or the product descriptions in your online store, your copy needs to be clear, persuasive, and aligned with your brand's voice.

Crafting a Clear Message

One of the most important rules for writing website copy is clarity. When visitors land on your website, they should immediately understand what your business does, what makes you different, and what you want them to do next. In short, your message should be simple and direct.

Key Questions to Answer in Your Copy:

Who are you? Clearly state what your business does.

What do you offer? Explain the products or services you provide.

Why should visitors care? Highlight the benefits and value you bring to customers.

What should visitors do next? Include clear calls to action (CTAs), such as "Contact Us," "Learn More," or "Shop Now."

Understanding Your Audience

Great website copy speaks directly to your audience. The tone, style, and level of detail in your writing should align with the preferences of your target market. For instance, a legal firm will use a more formal, authoritative tone, while a children's clothing store might adopt a friendly, playful voice.

Steps to Tailor Copy to Your Audience:

Identify Your Audience: Who is visiting your site? Are they professionals, families, students, or retirees? Understanding your audience helps shape the tone and language of your copy.

Address Their Pain Points: Consider the problems your target audience faces and how your products or services solve those problems. This helps make your copy relevant and persuasive.

Use Language They Understand: Avoid jargon or overly technical language that may confuse readers. Instead, write in a way that's clear and accessible to your audience.

Creating Engaging Headlines and Calls to Action

Your headlines and calls to action (CTAs) are the most important pieces of copy on your website. They grab visitors' attention and guide them to take action, so it's crucial to get them right.

Crafting Effective Headlines:

Be Clear and Specific: Your headline should immediately tell visitors what the page is about. For example, "Affordable Landscaping Services for Your Home" is more effective than "Welcome to Our Website."

Incorporate Keywords: If possible, include relevant keywords in your headlines to help with SEO and ensure search engines understand what your page is about.
Writing Strong Calls to Action (CTAs):

Be Direct: Use action-oriented language that tells visitors exactly what to do. For example, "Shop Now," "Get a Free Quote," or "Contact Us Today."

Create a Sense of Urgency: CTAs that suggest urgency or exclusivity can be more persuasive, such as "Limited Time Offer" or "Start Your Free Trial Today."

Make CTAs Visually Stand Out: Use contrasting colors or bold fonts for CTA buttons to make them stand out and easy to find.

Visual Content: Using Images and Videos Effectively

While written copy is important, visual content—like images and videos—plays an equally critical role in engaging visitors and telling your brand's story. High-quality visuals not only make your website more visually appealing, but they also help visitors connect with your business on an emotional level.

Importance of High-Quality Images

Images can greatly influence how visitors perceive your brand. Poor quality or irrelevant images can make your website look unprofessional, while high-quality images can build trust and make your site more engaging.

Tips for Using Images:

Use Original Images: Whenever possible, use original images of your products, team, or workplace. This adds authenticity and helps visitors connect with your brand on a personal level.

Invest in Professional Photography: If you don't have the resources to take your own photos, consider hiring a professional photographer. High-quality images are an investment that can pay off by making your website look polished and credible.

Optimize for Web: Large images can slow down your website's load time, which can hurt both user experience and SEO. Use compressed image files to ensure your site loads quickly.

Examples of Effective Image Use:

Product Pages: Use multiple high-quality images to showcase different angles or features of your products.

Team Pages: Include professional head shots of your team members to make your business feel more approachable and trustworthy.

Home and About Pages: Use images that reflect your brand's identity—whether that's images of your storefront, products, or services in action.

The Power of Video

Video content has become increasingly popular on websites because it's engaging and can convey information quickly. Whether it's a product demonstration, customer testimonial, or a behind-the scenes look at your business, video can help you connect with your audience in a more dynamic way.

Types of Videos to Consider:

Explainer Videos: Use short, engaging videos to explain your products or services. These are especially useful for complex offerings that require more explanation.

Customer Testimonials: Video testimonials from satisfied customers can build trust and serve as powerful social proof.

Behind-the-Scenes Videos: Show your audience how your products are made or introduce your team. This humanizes your brand and makes it more relatable.

Best Practices for Using Video:

Keep It Short: Aim for videos that are one to three minutes long. Visitors are less likely to engage with longer videos.

Optimize for Loading Speed: Like images, video files can slow down your site. Host your videos on platforms like YouTube or Vimeo and embed them on your website to keep your site fast.

Include Subtitles: Many users watch videos without sound, especially on mobile. Including subtitles ensures your message is clear even without audio.

Search Engine Optimization (SEO): Writing for Both People and Search Engines

When creating content for your website, it's important to consider SEO. Search Engine Optimization is the practice of optimizing your website's content so that it ranks higher on search engine results pages (SERPs), making it easier for people to find your site.

Keyword Research and Usage

One of the foundational elements of SEO is choosing the right keywords—the search terms that potential customers are using to find businesses like yours. Incorporating these keywords into your website's content can help improve your ranking on Google and other search engines.

Steps to Conduct Keyword Research:

Identify Core Keywords: Think about the main products or services your business offers. These should be the primary keywords you want to rank for.

Use Keyword Tools: Tools like Google Keyword Planner, Moz, or Ubersuggest can help you identify relevant keywords and estimate their search volume.

Incorporate Keywords Naturally: Once you've identified your target keywords, include them in your website's headings, copy, meta descriptions, and image alt text. However, be sure to use them naturally—Google penalizes sites that stuff too many keywords into their content.

Optimizing Your Content for SEO

Beyond keywords, there are several other ways to optimize your content for search engines while keeping it engaging for visitors.

Best Practices for SEO Optimization:

Use Headings and Subheadings: Organize your content with clear headings (H1, H2, H3) to make it easier for both users and search engines to navigate your site. Headings also help search engines understand the structure of your content.

Write Meta Descriptions: A meta description is the short blurb that appears under your website link in search engine results. It should be a concise, compelling summary of the page's content that includes your main keywords.

Optimize Image Alt Text: Search engines can't "see" images, so they rely on alt text to understand what an image is about. Be sure to include descriptive, keyword-rich alt text for all of your images.

Internal and External Links: Linking to other pages on your site (internal links) and to reputable external websites can help improve your SEO. Internal links also encourage visitors to explore more of your website, while external links help search engines see your content as credible.

Keeping Content Fresh and Relevant

It's not enough to create great content once and then leave it untouched. Regularly updating your content helps keep your website relevant, both for visitors and search engines. Fresh content signals to search engines that your site is active, which can improve your rankings.

Creating a Content Schedule

To keep your content fresh, consider creating a content calendar. This schedule can help you plan when to update pages, publish new blog posts, or refresh product descriptions.

How to Maintain a Content Schedule:

Regular Blog Posts: If you have a blog, aim to publish new posts on a regular basis (e.g., once a week or once a month). Blog posts provide fresh content and give you opportunities to target additional keywords.

Update Key Pages: Regularly review key pages, such as your homepage or product pages, to ensure the content is still accurate and relevant.

Seasonal or Promotional Updates: Update your content to reflect seasonal promotions or special offers. This not only attracts visitors but also keeps your site dynamic.

Creating Content That Engages and Converts

Content is the heart of your website—it's what draws visitors in, keeps them engaged, and ultimately drives them to take action. By focusing on clear, well-written copy, high-quality visuals, and SEO optimization, you can create a website that not only looks great but also delivers results.

By utilizing and putting emphasis on SEO services, you can ensure your website's content is tailored to your business needs, optimized for search engines, and designed to engage visitors from the moment they land on your site.

SEO and Digital Marketing

The Importance of SEO and Digital Marketing

Having a well-designed website is only half the battle—your website also needs to be discoverable. This is where Search Engine Optimization (SEO) and digital marketing come into play. SEO ensures that your site ranks higher in search engine results pages (SERPs), driving organic (unpaid) traffic to your website, while digital marketing focuses on promoting your website and brand through various online channels like social media, email marketing, and paid advertising.

In this chapter, we'll cover the basics of SEO, from keyword research to on-page optimization, and explore key digital marketing strategies that small business owners can implement to increase visibility, attract more visitors, and convert those visitors into loyal customers.

What Is SEO and Why Does It Matter?

SEO, or Search Engine Optimization, is the process of improving your website so that it ranks higher on search engines like Google, Bing, and Yahoo. Higher rankings mean more visibility, which leads to more website traffic and ultimately, more customers.

How Search Engines Work

Search engines use complex algorithms to determine which websites should appear at the top of search results. These algorithms take into account several factors, including the relevance of your content, the quality of your website, and how well it answers the searcher's query.

Key SEO Factors Search Engines Consider:

Keywords: Search engines look for keywords that match what users are searching for. Incorporating relevant keywords into your content helps search engines understand what your page is about.

Website Speed and Performance: Websites that load quickly and are optimized for mobile devices tend to rank higher.

Backlinks: These are links from other reputable websites that point to your website. The more high-quality backlinks you have, the more credible your site appears to search engines.

User Engagement: Metrics like time spent on your site, click-through rates, and bounce rates can indicate to search engines whether users find your content valuable.

The Benefits of SEO for Small Businesses

SEO offers several key benefits for small business owners:
Increased Visibility: Ranking higher in search results means more people will discover your website, increasing brand awareness.

Cost-Effective Traffic: SEO attracts organic traffic, which is free compared to paid advertising. While it may take time to see results, the long-term benefits of ranking well can reduce your need for expensive ads.

Targeted Traffic: SEO allows you to target specific keywords related to your business, attracting users who are actively searching for the products or services you offer.

Keyword Research: The Foundation of SEO

Keyword research is the process of finding and analyzing the search terms that your potential customers are using. Choosing the right keywords is crucial for driving the right kind of traffic to your website.

Types of Keywords

There are several types of keywords you should be aware of when optimizing your website:

Short-Tail Keywords: These are broad, one- or two-word phrases like "shoes" or "plumbing." While they get a lot of traffic, they are highly competitive and may not always bring in qualified leads.

Long-Tail Keywords: These are more specific phrases, such as “women’s running shoes in Miami” or “emergency plumbing services near me.” Long-tail keywords tend to have less competition and attract visitors who are closer to making a purchase.

Example:

For a local bakery in New York, instead of targeting a broad keyword like “cakes,” you could target “custom birthday cakes in New York City” to attract customers specifically looking for your services.

Conducting Keyword Research

To find the best keywords for your business, follow these steps:

Step 1: Brainstorm Keywords

Start by brainstorming a list of words and phrases that are relevant to your business. Think about what potential customers might type into a search engine when looking for your products or services.

Step 2: Use Keyword Tools

There are several tools you can use to research keywords, including:

Google Keyword Planner: A free tool that shows search volume and competition for specific keywords.

SEMrush or Ahrefs: Paid tools that offer more in-depth keyword analysis, including how difficult it is to rank for certain terms and what your competitors are ranking for.

Step 3: Analyze Search Volume and Competition

For each keyword, you'll want to assess the search volume (how many people are searching for that term) and the competition (how difficult it will be to rank for it). Aim for keywords that have a healthy balance of high search volume and low to medium competition.

Step 4: Focus on Local Keywords

If your business serves a specific geographic area, prioritize local keywords like "best Italian restaurant in Austin" or "roof repair in Chicago." These keywords help attract nearby customers who are ready to do business.

On-Page SEO: Optimizing Your Website Content

Once you've identified the right keywords, the next step is to optimize your website's content. This process is known as on-page SEO and involves making sure your content and website structure are search engine-friendly.

Optimizing Meta Tags

Meta tags are pieces of HTML code that help search engines understand the content of your website. There are two main types of meta tags you should focus on:

Title Tag: This is the clickable headline that appears in search engine results. It should be concise (50-60 characters), include your main keyword, and be written in a way that encourages users to click.

Meta Description: This is the short description that appears below the title tag in search results. While it doesn't directly impact rankings, it can influence whether users click on your result. Make sure it includes your primary keyword and clearly describes the page's content in about 150-160 characters.

Optimizing Headings (H1, H2, H3)

Headings help organize your content and make it easier for both users and search engines to understand. Use the following guidelines for your headings:

H1 Tag: Each page should have one main heading (H1) that includes the primary keyword and clearly describes the page's content. For example, if your page is about dog grooming services, the H1 might be "Affordable Dog Grooming in Los Angeles."

H2 and H3 Tags: These subheadings break up your content and make it more readable. Use relevant keywords in these headings to help search engines understand the page's structure.

Image Optimization

Search engines can't "see" images, so it's important to include alt text that describes the image. Alt text not only improves SEO but also makes your site more accessible for people with visual impairments.

Alt Text: Be descriptive and include relevant keywords. For example, instead of "image001.jpg," use "affordable dog grooming services in Los Angeles."

Internal and External Linking

Linking to other pages on your website (internal links) and linking to credible external websites (external links) both help search engines understand the structure of your content and improve your SEO.

Internal Links: Guide users to other relevant pages on your site, such as linking from a blog post to a product page. This helps keep visitors on your site longer and encourages them to explore more content.

External Links: Link to authoritative, relevant sources when appropriate. For example, if you reference a study or article, link to it. This signals to search engines that your content is credible and well-researched.

Off-Page SEO: Building Backlinks and Social Signals

While on-page SEO focuses on optimizing your website, off-page SEO involves actions taken outside your website to boost its authority and relevance. The most important aspect of off-page SEO is link building, or earning backlinks from other reputable websites.

What Are Backlinks?

Backlinks are links from other websites that point to your website. Search engines see backlinks as a signal that your site is trustworthy and valuable, so the more high-quality backlinks you have, the better your site will rank.

How to Earn Backlinks:

Create High-Quality Content: The best way to earn backlinks is by creating valuable content that other sites want to link to, such as informative blog posts, infographics, or research reports.

Guest Blogging: Offer to write guest posts for reputable blogs in your industry, and include a link back to your website in your author bio.

Reach Out to Industry Websites: If you've published a particularly valuable piece of content, reach out to industry websites and suggest that they link to it as a helpful resource.

Social Media Signals

While social media links don't directly affect SEO rankings, they can drive traffic to your website and increase its visibility. When users share your content on platforms like Facebook,

Twitter, or Instagram, it increases the likelihood of earning backlinks and boosts your brand's online presence.

Digital Marketing: Promoting Your Website Online

In addition to SEO, you'll need a robust digital marketing strategy to drive traffic to your website. Digital marketing includes tactics like social media marketing, email marketing, and online advertising to reach a broader audience and attract potential customers.

Social Media Marketing

Social media marketing allows you to connect with your audience on platforms like Facebook, Instagram, LinkedIn, and Twitter. By sharing valuable content and engaging with followers, you can build brand awareness and drive traffic to your website.

Key Social Media Strategies:

Consistent Posting Schedule: Regularly post content that is relevant to your audience, such as tips, product updates, or industry news.

Engagement: Respond to comments and messages, and encourage followers to share your content. The more engagement your posts get, the more visible your brand becomes.

Paid Ads: Social media platforms offer targeted advertising options. For example, Facebook ads allow you to reach users based on demographics, interests, and behaviors, making it easier to target your ideal customers.

Email Marketing

Email marketing is one of the most cost-effective ways to reach your audience directly. With the right email strategy, you can nurture relationships with potential customers, promote new products or services, and encourage repeat business.

Building an Email List:

Offer Incentives: Encourage website visitors to sign up for your email list by offering a discount, free guide, or exclusive content.

Regular Newsletters: Send regular newsletters with valuable content, updates, and promotions to keep your audience engaged and informed.

Online Advertising (PPC)

If you want to see quicker results, Pay-Per-Click (PPC) advertising can be an effective way to drive traffic to your website. With PPC ads, you bid on keywords, and your ad appears at the top of search results when users search for those terms. You only pay when someone clicks on your ad.

Platforms to Consider:

Google Ads: The most popular PPC platform, allowing you to target specific keywords and appear in search results.

Social Media Ads: Platforms like Facebook and Instagram allow you to run targeted ads based on user demographics and interests.

Conclusion: Boosting Your Website's Visibility with SEO and Digital Marketing

SEO and digital marketing are powerful tools for driving traffic to your website and growing your business online. By optimizing your site for search engines, building backlinks, and using digital marketing strategies like social media, email marketing, and online advertising, you can reach more potential customers and increase your online presence.

Website Maintenance and Security

Why Website Maintenance and Security Matter

A website isn't a set-it-and-forget-it project. Once your website is live, it requires ongoing maintenance and updates to ensure it continues to function properly, remains secure, and delivers a great user experience. Neglecting these tasks can lead to slow load times, security breaches, outdated content, and even a loss of customers.

In this chapter, we'll explore the essential maintenance tasks every small business owner should perform regularly, the security measures needed to protect your website from cyber threats, and how to keep your site running smoothly over time. With the right maintenance plan, you can ensure your website remains an asset to your business instead of becoming a liability.

Regular Website Maintenance: Keeping Your Site Running Smoothly

Just like a car needs regular servicing, your website requires routine maintenance to stay in top shape. Failing to maintain your website can lead to broken links, slow performance, and even downtime that frustrates users and hurts your SEO rankings.

Software Updates

If your website is built on a content management system (CMS) like WordPress, Joomla, or Drupal, it's important to keep your software updated. These systems frequently release updates to improve functionality, fix bugs, and patch security vulnerabilities.

Key Components to Update:

CMS Core Updates: Always update to the latest version of your CMS to ensure your website stays secure and compatible with new features.

Plugins and Themes: Many websites rely on plugins or themes to add functionality and design elements. These also need regular updates to stay secure and fully functional.

Pro Tip: Many updates are automated or can be set to run automatically, but it's important to check for any custom changes or compatibility issues before performing an update.

Backing Up Your Website:

Regular backups are critical to protecting your website from data loss. If your site is hacked, or if a plugin update causes issues, a backup allows you to restore your website to a previous version quickly.

Types of Website Backups:

Full Backups: These backups save everything on your website, from images and content to plugins and databases. A full backup is ideal for restoring your entire website after a significant problem.

Incremental Backups: These backups save only the changes made since the last backup, reducing the amount of data stored and the time it takes to perform a backup.

Backup Best Practices:

Frequency: Backups should be performed regularly. For active websites (like e-commerce sites or blogs), daily backups are recommended. Less active websites may only need weekly or monthly backups.

Offsite Storage: Always store backups in a secure location separate from your website's server, such as cloud storage or an external hard drive.

Pro Tip: Many providers offer automated backup services, ensuring your website's data is safely stored and can be restored quickly in case of an emergency.

Testing Website Functionality

Over time, certain elements of your website may stop working correctly due to software updates, plugin conflicts, or other issues. Regularly testing your website's functionality helps catch and fix problems before they affect your users.

What to Test:

Forms and Contact Pages: Make sure all contact forms, inquiry forms, and booking systems are working properly and that submissions are being received.

E-commerce Functionality: If you run an online store, check that the shopping cart, checkout, and payment processes are working smoothly.

Links: Ensure all internal and external links on your website are still valid and aren't leading to 404 errors (broken pages).

Monitoring Website Speed and Performance

Website speed plays a major role in user experience and SEO rankings. If your site takes too long to load, visitors may leave before the page even appears, which can hurt your conversion rates.

How to Monitor Website Speed:

Google PageSpeed Insights: This free tool analyzes your website's speed and provides suggestions for improving performance on both desktop and mobile devices.

GTmetrix or Pingdom: These tools offer more detailed performance reports and highlight areas where your site may be lagging, such as large images or too many server requests.

Ways to Improve Site Speed:

Optimize Images: Compress images and use modern file formats like WebP to reduce load times.

Leverage Browser Caching: Browser caching stores some of your website's resources in users' browsers so that they don't need to be reloaded every time a visitor returns to your site.

Minimize JavaScript and CSS Files: Combining and minimizing these files can reduce the number of requests your server needs to process, speeding up load times.

Security Essentials: Protecting Your Website from Cyber Threats

Website security is critical for protecting your business and your customers. Cyber threats like hacking, malware, and data breaches can damage your reputation, lead to financial losses, and even cause legal issues. Implementing the right security measures can prevent these problems and keep your website safe.

The Importance of SSL Certificates

One of the most fundamental security measures for any website is an SSL certificate. SSL (Secure Sockets Layer) encrypts the data transferred between your website and its visitors, protecting sensitive information like credit card details and passwords from being intercepted by hackers.

Benefits of SSL:

Data Encryption: SSL ensures that all data transferred between the visitor's browser and your website is encrypted, keeping it secure from cybercriminals.

Trust and Credibility: Visitors feel more confident when they see the padlock icon and "https://" in the browser bar, indicating that your site is secure.

SEO Boost: Google considers SSL a ranking factor, meaning secure websites are more likely to rank higher in search results.

Pro Tip: Most hosting providers, offer free or affordable SSL certificates with their hosting plans.

Preventing Hacking and Malware

Even small business websites are targets for hackers, who can exploit vulnerabilities to steal data, plant malware, or deface your site. Here are some key steps you can take to prevent hacking and malware attacks:

Keep Software Updated

Outdated software is one of the most common ways hackers gain access to websites. Always update your CMS, plugins, themes, and any other software to the latest versions to patch security holes. Use Strong Passwords and Two-Factor Authentication (2FA)

Ensure that all admin accounts use strong, unique passwords that include a combination of letters, numbers, and special characters. Enabling two-factor authentication (2FA) adds an extra layer of security, requiring users to enter a code sent to their phone or email when logging in.

Install a Web Application Firewall (WAF)

A web application firewall (WAF) acts as a barrier between your website and potential threats by filtering and monitoring incoming traffic for malicious activity.

Scan for Malware Regularly

Use security tools to regularly scan your website for malware. Many website security plugins (like Wordfence for WordPress) automatically scan for malware and alert you if anything suspicious is detected.

Protecting Against DDoS Attacks

Distributed Denial of Service (DDoS) attacks are a type of cyberattack where hackers flood your website with massive amounts of traffic, overwhelming your server and causing your site to crash. While no website is immune to DDoS attacks, there are measures you can take to reduce the risk.

How to Prevent DDoS Attacks:

Use a Content Delivery Network (CDN): CDNs distribute your website's content across multiple servers worldwide, reducing the load on your main server and mitigating the impact of DDoS attacks.

Monitor Traffic Spikes: Set up monitoring tools to alert you if there's an unusual spike in traffic that could indicate a potential DDoS attack.

Enable DDoS Protection from Your Hosting Provider: Many hosting providers offer DDoS protection as part of their services.

Keeping Your Website Secure

While no security system is foolproof, following best practices can significantly reduce your risk of a security breach.

Limit Access to Your Website

Only give access to people who need it. If you have employees or contractors managing different parts of your website, ensure they have their own login credentials with the appropriate permissions. Avoid sharing a single admin account across multiple users, as this increases the risk of security breaches.

Regularly Change Passwords

Periodically changing your website's admin passwords reduces the risk of unauthorized access, especially if an old password has been compromised in a data breach. Use a password manager to generate and store complex passwords securely.

Conduct Security Audits

A security audit involves reviewing your website's security settings, software, and configurations to identify vulnerabilities. Regular audits can catch security gaps before they become major issues.

What to Include in a Security Audit:

Review access logs to identify any suspicious activity.

Ensure all software is up to date and check for any known vulnerabilities in plugins or themes.

Test backup restoration to ensure you can recover your website quickly in the event of a breach or malware infection.

Pro Tip: A great hosing provider will offer security audits, and maintenance plans that regularly monitor your website's security, ensuring it's protected from emerging threats.

Maintaining and Protecting Your Website

Website maintenance and security are essential for keeping your business's online presence strong and reliable. By performing regular updates, backups, and security checks, you can prevent technical issues, improve your website's performance, and protect it from cyber threats.

Success: Analytics and Performance Tracking

The Importance of Website Analytics

Building and launching your website is only the first step in establishing your online presence. To ensure your website is truly effective, you need to measure its performance and understand how visitors interact with your site. This is where website analytics comes into play.

By tracking and analyzing website data, you can identify what's working, what needs improvement, and how you can enhance the user experience to drive more traffic, conversions, and sales. In this chapter, we'll explore the most important metrics to track, how to use tools like Google Analytics, and ways to optimize your website based on the data you collect.

Understanding Website Metrics

When it comes to tracking the success of your website, there are several key performance indicators (KPIs) that give you insight into how well your site is performing. These metrics can help you assess user behavior, identify areas for improvement, and optimize your website for better results.

Traffic Metrics

Traffic metrics show how many people are visiting your website, where they're coming from, and how they navigate through your site. These metrics provide an overview of your website's reach and help you understand the effectiveness of your marketing efforts.

Key Traffic Metrics:

Total Visitors: This metric shows how many unique users have visited your website within a specific time period (e.g., daily, weekly, or monthly). Monitoring traffic trends helps you understand if your site's popularity is growing or declining.

Traffic Sources: Traffic can come from various sources, such as organic search (from search engines), direct traffic (users typing your URL), referrals (links from other websites), or social media. Understanding which sources are driving the most traffic allows you to focus on what's working.

New vs. Returning Visitors: This metric shows the ratio of first-time visitors to repeat visitors. A higher percentage of returning visitors may indicate that your content is engaging and users are coming back for more.

Example:

If you run a local coffee shop and notice a spike in traffic from social media, it could mean that a recent post or promotion resonated well with your audience, helping you drive more visitors to your site.

User Behavior Metrics

User behavior metrics help you understand how visitors interact with your website. By tracking how long visitors stay on your site, which pages they view, and whether they take desired actions, you can identify areas where users are engaging or where they might be dropping off.

Key User Behavior Metrics:

Bounce Rate: This is the percentage of visitors who leave your website after viewing only one page. A high bounce rate can indicate that visitors aren't finding what they're looking for, or that your landing pages need improvement.

Session Duration: The average time visitors spend on your website. Longer session durations generally indicate that users are finding your content engaging and are exploring multiple pages.

Pages per Session: This shows how many pages a visitor typically views during a single session. It gives insight into how well your site's content encourages users to explore.

Exit Rate: The percentage of visitors who leave your site from a specific page. Monitoring exit rates helps you identify potential problem areas, such as pages with confusing content or broken links.

Example:

If a visitor lands on your product page and exits immediately (high bounce rate), you might need to improve the product description, add more images, or simplify the checkout process.

Conversion Metrics

Conversions are the actions you want visitors to take on your website, such as filling out a contact form, signing up for a newsletter, or completing a purchase. Conversion metrics help you track how successful your site is at turning visitors into leads or customers.

Key Conversion Metrics:

Conversion Rate: The percentage of visitors who complete a specific action (e.g., making a purchase or submitting a form). Conversion rates help you assess how well your site is meeting your business goals.

Goal Completions: Most analytics tools allow you to set up specific goals, such as completing a checkout or booking an appointment. Tracking these goal completions helps you measure success against your objectives.

Cart Abandonment Rate (for e-commerce): The percentage of users who add items to their cart but leave the site without completing a purchase. A high cart abandonment rate may signal issues with the checkout process, such as complicated forms or high shipping costs.

Example:

For an online clothing store, a conversion could be a visitor completing a purchase. Tracking conversion rates can help you identify whether your product pages, CTA buttons, or checkout process need optimization.

Setting Up Google Analytics

One of the most powerful and widely used tools for tracking website performance is Google Analytics. It provides comprehensive data on your website's traffic, user behavior, and conversions, allowing you to make informed decisions based on real-time data.

How to Set Up Google Analytics

Setting up Google Analytics is straightforward, and it's free to use. Follow these steps to get started:

Step 1: Create a Google Analytics Account

Sign up for a Google Analytics account at analytics.google.com. If you already have a Google account, you can use it to sign in.

Step 2: Set Up a Property for Your Website

After signing in, click "Set up for free" and enter your website's details. Google will generate a Tracking ID or Measurement ID that you'll need to add to your website.

Step 3: Add the Tracking Code to Your Website

For most websites, you'll need to add the tracking code to the header of every page. If you're using a CMS like WordPress, there are plugins (like Google Site Kit) that make adding the tracking code easier.

Step 4: Verify the Installation

Once the tracking code is installed, you can verify that it's working by checking the "Real-Time" section in Google Analytics. If data is being collected, you'll see live traffic data for your site.

Navigating Google Analytics

Once your account is set up, Google Analytics provides a wealth of data. Here are some key sections to focus on:

Audience Overview

This section gives you an overview of your website's visitors, including metrics like total visitors, sessions, and demographics. It helps you understand who is visiting your site and how frequently.

Acquisition Overview

This section shows where your website traffic is coming from. It breaks down traffic sources, such as organic search, paid ads, social media, and referrals. You can use this information to evaluate which channels are driving the most visitors.

Behavior Overview

This section provides insights into how visitors interact with your website, including page views, bounce rates, and average session duration. Use this data to identify which pages are performing well and which need improvement.

Conversion Overview

In the conversion section, you can track goal completions and conversion rates. Set up specific goals, such as a form submission or purchase, and monitor how well your site is converting visitors into leads or customers.

Using Data to Improve Website Performance

Once you've set up analytics and understand your website's metrics, the next step is to use the data to improve your website. By identifying patterns in user behavior, you can make informed changes that lead to better user experiences and higher conversion rates.

Reducing Bounce Rates

A high bounce rate can indicate that visitors aren't finding what they expected on your website or that the page is slow to load. Here are some strategies to reduce bounce rates:

Optimize Page Load Times: A slow-loading website is one of the top reasons visitors leave a site. Use tools like Google PageSpeed Insights to identify and fix performance issues.

Improve Content Relevance: Ensure that your content matches what visitors are expecting based on the keywords or search queries that brought them to your site.

Enhance Visual Appeal: A visually engaging page can keep visitors interested. Invest in high-quality images, graphics, and a clean layout that encourages users to explore more.

Increasing Conversion Rates

If you're seeing low conversion rates, there are several ways to improve the user experience and encourage more visitors to take action:

Simplify CTAs: Ensure that your call-to-action (CTA) buttons are clear, concise, and easy to find. Use action-oriented language like "Sign Up Now" or "Get Started."

Streamline the Checkout Process: For e-commerce sites, a complicated checkout process can lead to abandoned carts. Simplify forms, offer guest checkout options, and make sure payment methods are easy to use.

A/B Testing: Run A/B tests to compare different versions of key pages (such as landing pages or product pages) and see which one performs better. Test different headlines, images, or CTA buttons to find the most effective combinations.

Optimizing Pages with High Exit Rates

If certain pages have a high exit rate, it may indicate that visitors are losing interest or encountering problems. Here are a few ways to address this issue:

Check for Broken Links: Ensure that all links are working properly and that users aren't encountering 404 errors.

Improve Content Quality: Review the content on high-exit pages to ensure it's engaging and provides value. Consider adding more detailed information, images, or videos.

Add Internal Links: Encourage visitors to stay on your site by including internal links to other relevant pages, such as blog posts, products, or service offerings.

Heatmaps in Understanding User Behavior

In addition to standard analytics tools, heatmaps can provide valuable insights into how visitors are interacting with your website. Heatmaps show where users are clicking, scrolling, and spending the most time, allowing you to see which areas of your website are getting the most attention.

How Heatmaps Work

Heatmaps use color gradients to show areas of high and low activity on your website. Areas that are clicked or scrolled through frequently will appear in red or orange, while less active areas will appear in blue or green.

Benefits of Using Heatmaps:

Identify Popular Elements: See which buttons, links, or sections are getting the most attention. This can help you place important CTAs in high-traffic areas.

Understand User Behavior: By analyzing where users are clicking or spending the most time, you can optimize your layout to guide them toward conversion points.

Example:

If a heatmap shows that users are frequently clicking on an image that isn't a link, you may want to turn that image into a clickable CTA to improve the user experience.

Using Data to Drive Website Success

Measuring the success of your website is essential for continuous improvement. By tracking key performance indicators (KPIs) like traffic, user behavior, and conversion rates, you can identify what's working and where there's room for optimization. With the right tools—like Google Analytics and heatmaps—you can gather the insights needed to make data-driven decisions that improve user experience, increase conversions, and ultimately, grow your business.

Scaling and Growing Your Online Presence

Why Growth Requires Adaptation

As your business grows, your website must grow with it. Whether you're expanding your product line, entering new markets, or increasing traffic through marketing efforts, your website needs to be able to handle the demands of a larger, more dynamic business. This means adapting your website with new features, optimizing it for higher traffic, and scaling your digital marketing efforts.

In this chapter, we'll explore how to scale your website and online presence effectively. We'll cover adding new features, upgrading infrastructure, expanding your digital marketing strategy, and integrating advanced tools to support long-term growth.

Adding New Features: Expanding Functionality

As your business evolves, you may need to add new features to your website to better serve customers and streamline operations. Whether you're adding an e-commerce platform, customer portals, or integrating marketing automation, it's important to think ahead and plan for scalability.

E-Commerce Integration

If your business starts as a service-based company or brick-and-mortar store, you may eventually decide to sell products online. Adding e-commerce functionality allows you to reach a broader audience and create an additional revenue stream.

Key E-Commerce Features to Consider:

Product Pages: Detailed product descriptions, high-quality images, and easy-to-use filtering options.

Shopping Cart and Checkout: A seamless and secure shopping cart and checkout experience that supports multiple payment methods (e.g., credit cards, PayPal, Apple Pay).

Inventory Management: Real-time inventory tracking to manage stock levels and alert customers when products are low or out of stock.

Shipping Integration: Automated shipping cost calculations and tracking options to provide a smooth post-purchase experience for customers.

Example:

A local clothing boutique may initially rely on foot traffic but could scale by adding an e-commerce store, allowing customers to shop online from anywhere. This expansion requires integration of a payment gateway, shipping solutions, and product management.

Customer Portals and Self-Service Features

As your business grows, so does the demand for more efficient customer service. Adding customer portals or self-service features can help reduce the burden on your team by allowing customers to manage their accounts, access personalized content, and complete transactions on their own.

Features to Consider:

Account Management: Allow customers to create accounts, view order history, track shipments, or manage subscriptions.

Appointment Scheduling: If you offer services, integrate a scheduling tool so customers can book appointments or consultations online.

Customer Support Tools: Add features like live chat, a knowledge base, or a ticketing system to improve customer service.

Example:

A dental office that originally handled all appointment bookings via phone could scale by adding a self-service online booking system. This would allow patients to choose available time slots at their convenience, freeing up office staff for other tasks.

Blog and Content Expansion

Expanding your website's content—through blogs, articles, or educational resources—can help establish your business as an authority in your industry and drive more organic traffic. As you grow, creating a regular content schedule and producing high-quality blog posts or case studies can help attract more visitors and improve your SEO rankings.

Benefits of Expanding Content:

Increase Traffic: Regularly updated blogs and articles can help you rank for more keywords, leading to increased organic search traffic.

Establish Authority: By providing valuable, informative content, you can position yourself as a thought leader in your industry and build trust with your audience.

Lead Generation: Offer downloadable resources (e.g., eBooks, guides, white papers) in exchange for contact information to build your email list.

Example:

A financial adviser could start publishing weekly blog posts offering tips on saving for retirement or investing in the stock market. This type of content attracts traffic and positions the adviser as a reliable expert in their field.

Preparing for Increased Traffic

As your website gains more visitors, it's important to ensure that your infrastructure can handle the increased traffic without slowing down or crashing. This means upgrading your hosting plan, optimizing your site's speed, and ensuring that your server resources scale with your business.

Upgrading Your Hosting Plan

As your website grows, a basic shared hosting plan may no longer be sufficient. Upgrading to a VPS (Virtual Private Server) or dedicated hosting plan can offer more resources, improved performance, and better security for high-traffic websites.

Hosting Options:

VPS Hosting: A step up from shared hosting, VPS hosting provides more dedicated resources and flexibility, making it ideal for growing businesses that expect moderate to high traffic.

Dedicated Hosting: For larger websites with heavy traffic, dedicated hosting provides the highest level of performance and control, as your website has an entire server to itself.

Example:

A growing online retailer with increasing traffic may need to move from shared hosting to VPS hosting to ensure the site remains fast and responsive, especially during peak shopping seasons like holidays or sales events.

Optimizing Website Speed for Growth

As your website grows, maintaining fast load times becomes critical.

Slow websites can lead to higher bounce rates, lower SEO rankings, and poor user experience. To ensure your site remains fast as traffic increases, it's important to optimize for speed.

Ways to Improve Website Speed:

Content Delivery Network (CDN): CDNs distribute your website's content across multiple servers worldwide, ensuring faster load times for visitors, regardless of their location.

Image Optimization: Compress images and use modern formats (like WebP) to reduce file sizes without compromising quality.

Caching: Implement browser and server-side caching to speed up page load times by storing static files that don't need to be reloaded every time a user visits your site.

Example:

A travel agency that starts receiving high volumes of traffic from international visitors can use a CDN to ensure that their website loads quickly for users in different regions, improving both user experience and SEO.

Expanding Your Online Reach

As your business grows, so should your digital marketing efforts. Scaling your digital marketing means reaching new audiences, driving more traffic, and increasing conversions. This can be achieved through advanced marketing tactics like targeted advertising, marketing automation, and lead nurturing.

Expanding Social Media Marketing

Social media marketing is a powerful tool for engaging with your audience and driving traffic to your website. As your business grows, it's important to scale your social media efforts by increasing your posting frequency, exploring new platforms, and using paid advertising.

Advanced Social Media Strategies:

Paid Social Ads: Platforms like Facebook, Instagram, and LinkedIn offer targeted advertising options that allow you to reach specific audiences based on demographics, interests, and behaviors.

Influencer Collaborations: Partnering with influencers in your industry can help you reach a broader audience and build credibility with potential customers.

Video Content: As video becomes increasingly popular, creating short, engaging videos for platforms like Instagram Reels or YouTube can help you capture more attention.

Example:

A local restaurant might initially rely on organic social media posts, but as they grow, they could start using paid ads to target specific demographics (e.g., foodies or locals in the area) or partner with local food bloggers for influencer collaborations.

Automating Email Marketing and Lead Nurturing

As your email list grows, marketing automation becomes essential for managing campaigns and nurturing leads. Automating your email marketing allows you to send personalized messages at scale, increasing engagement and conversions without requiring manual effort.

Email Automation Strategies:

Welcome Series: Automatically send new subscribers a series of emails introducing your brand, offering discounts, or highlighting popular products or services.

Abandoned Cart Emails: For e-commerce businesses, set up automated emails to remind customers who added items to their cart but didn't complete their purchase.

Lead Nurturing: Segment your email list based on customer behavior (e.g., past purchases or content downloads) and send targeted messages that guide them through the sales funnel.

Example:

An online fitness coach could use email automation to send new subscribers a welcome series, followed by workout tips or special offers for personal training services, depending on their interests.

Paid Search Advertising (PPC)

As your business scales, investing in Pay-Per-Click (PPC) advertising can help you reach a larger audience quickly. PPC campaigns, like those offered through Google Ads, allow you to bid on keywords and display ads at the top of search results for relevant queries.

Benefits of PPC Advertising:

Immediate Results: Unlike SEO, which takes time to show results, PPC ads can drive traffic to your site almost instantly.

Targeted Ads: PPC platforms allow you to target specific demographics, geographic areas, and even the time of day your ads appear.

Control Over Budget: You can set a daily or monthly budget, ensuring you don't overspend while still reaching potential customers.

Example:

A landscaping company that wants to expand its customer base could run PPC campaigns targeting local homeowners searching for terms like "lawn care services near me." PPC ads would help them show up at the top of search results for these high-intent keywords.

Tracking Key Performance Indicators (KPIs)

As you scale your website and marketing efforts, it's essential to track key performance indicators (KPIs) to measure growth and identify areas for improvement.

Key Metrics to Track as You Scale

As your business grows, the following KPIs will help you assess your website's performance and marketing success:

Website Traffic Growth: Track the total number of visitors over time, as well as traffic sources, to ensure that your marketing efforts are driving more users to your site.

Conversion Rates: Monitor conversion rates (such as form submissions, purchases, or email sign-ups) to ensure that increased traffic is leading to more sales or leads.

Customer Lifetime Value (CLV): Measure the average revenue generated from each customer over time. As you scale, CLV should increase if you're successfully retaining customers and encouraging repeat purchases.

Customer Acquisition Cost (CAC): Track how much you're spending to acquire each new customer. A healthy CAC should decrease as your marketing efforts become more efficient.

Example:

An online retailer scaling its business may track growth in monthly website visitors, conversion rates from paid ads, and the customer lifetime value to ensure their scaling efforts are driving profitability.

Preparing for Long-Term Growth

Scaling your online presence is an exciting part of business growth, but it requires strategic planning and the right tools. By adding new features, upgrading your website's infrastructure, expanding your digital marketing efforts, and tracking key performance indicators, you can ensure your business is ready to meet the challenges and opportunities that come with growth.

Case Studies or Success Stories Local Service Business – Sunrise Landscaping Background

Sunrise Landscaping is a small landscaping company based in Austin, Texas. For years, the company relied solely on word-of-mouth referrals and local flyers to get new clients. The business was performing well, but owner Daniel Kim knew he could grow the company if he had a stronger online presence. His goal was to create a website that could attract more local customers, showcase the quality of his work, and allow potential clients to book consultations online.
Challenges

Sunrise Landscaping faced several challenges in building its first website:

Limited Online Presence: With no website, the business had no way for potential customers to find it through online searches.

Lack of Digital Expertise: Daniel had no experience with web design or digital marketing and wasn't sure where to start.

Booking Inefficiencies: All client inquiries and bookings were handled via phone, which led to missed calls and lost opportunities.

Solution

After researching options, Daniel chose to work with a web hosting and development service that offered affordable, professional website design for small businesses. They built a simple, clean website for Sunrise Landscaping that included:

Service Pages: Detailed descriptions of landscaping services, including lawn care, garden design, and tree trimming, each optimized for local SEO keywords like "Austin landscaping services."

Online Booking System: A booking system was integrated, allowing visitors to schedule consultations directly through the website. This made the process more convenient for both the clients and the business.

Portfolio Section: A gallery showcasing before-and-after photos of landscaping projects, giving potential clients visual proof of the company's work quality.

Mobile-Friendly Design: With many customers searching for services on their phones, the site was built to be fully responsive on mobile devices.

Results

Within six months of launching the new website, Sunrise Landscaping saw a significant uptick in business:

40% increase in new client inquiries through the online booking system.

20% boost in revenue, driven by new clients who found the company via local Google searches.

Improved customer experience, as clients could now schedule consultations and view project examples online at any time, reducing the number of missed calls and inquiries.

With the help of a hosting provider than specializes in SEO optimization, the website now ranks on the first page of Google for local search terms like "landscaping services in Austin," making it easier for potential customers to find Sunrise Landscaping online.

E-Commerce – The Handmade Soap Company

Background

The Handmade Soap Company is an artisan soap business founded by Sarah Davis. Initially, Sarah sold her soaps at local farmers' markets and through Instagram, but she wanted to expand her reach and sell products to a national audience. Her goal was to create an e-commerce website that showcased her handmade products and allowed her to sell directly to customers online.

Challenges

As a solo entrepreneur, Sarah faced several challenges:

Manual Order Processing: Sarah managed orders via direct messages on Instagram, leading to a disorganized process and delays in fulfilling orders.

Limited Sales Reach: Without a professional website, she was missing out on potential customers who preferred shopping online.

Branding and Presentation: Sarah wanted to create a visually appealing website that reflected the natural and handcrafted nature of her products.

Solution

Sarah partnered with a local host to build a fully functional ecommerce website that met her needs:

Product Pages: The site featured individual product pages for each type of soap, with high-quality photos, detailed descriptions, and pricing. These pages were optimized for SEO with keywords like "natural handmade soap" and "organic skincare."

Integrated Shopping Cart and Payment Gateway: The new website included a shopping cart and checkout process, allowing customers to purchase products directly from the site using credit cards, PayPal, or Apple Pay.

Brand Storytelling: A dedicated "About Us" section highlighted Sarah's journey and the ethical, natural ingredients used in her soaps, helping to build trust with her audience.

Customer Reviews and Testimonials: The hosting company integrated a customer review system, allowing buyers to leave feedback, which helped build social proof and increase credibility.

Results

The Handmade Soap Company experienced rapid growth within a few months of launching the website:

300% increase in sales, with most new orders coming from out-of-state customers who discovered the website via online searches.

Significant time savings, as the automated shopping cart and order system replaced Sarah's manual order process, allowing her to focus on product creation and marketing.

Higher Average Order Value (AOV): With bundled product offers and discounts for bulk purchases, the average order value increased by 25%.

Thanks to the hosting company's SEO work, the website now ranks for key search terms like "handmade organic soap" and "natural soap gifts," bringing in organic traffic from across the U.S.

Local Retail Store – Smith's Bike Shop

Background

Smith's Bike Shop is a family-owned business in Portland, Oregon, that has been selling and repairing bikes for over 20 years. While the store had a loyal local customer base, owner Alex Smith wanted to modernize the business by creating a website that would allow customers to browse products online, book repair services, and view upcoming events like group rides and community workshops.

Challenges

The shop faced a few key challenges:

Outdated Marketing Methods: Smith's Bike Shop relied heavily on walk-in traffic and word-of-mouth marketing, making it difficult to attract new customers in the digital age.

No Online Booking: Customers could only schedule repairs by phone, leading to missed appointments and inefficient scheduling.

Lack of Online Presence: With no website, the shop had minimal online visibility, which was limiting its ability to attract younger, tech savvy customers.

Solution

Smith's Bike Shop worked with a managed website and hosting company to develop a website that provided a seamless experience for customers, both in-store and online:

Product Catalog: The website featured a catalog of available bikes, parts, and accessories, complete with detailed descriptions and pricing. While the store didn't sell products online initially, customers could browse items and call the store to place orders.

Service Booking System: The hosting company integrated an online booking system, allowing customers to schedule bike repairs or tuneups directly from the website.

Event Calendar: The site included an events page where customers could view and sign up for upcoming workshops, group rides, and community events.

Mobile Optimization: With many bike enthusiasts searching for services while on the go, The development company ensured the website was fully optimized for mobile devices.

Results

Smith's Bike Shop saw a significant improvement in both customer engagement and sales after launching its website:

30% increase in repair bookings, as customers now had the convenience of scheduling services online.

20% boost in in-store sales, driven by online visitors who browsed the product catalog before coming into the shop.

Expanded customer base, with younger, more tech-savvy customers discovering the shop through local Google searches.

The website now ranks well for local SEO keywords like "bike shop in Portland" and "bike repair services near me," helping the shop attract more online visibility and foot traffic.

Conclusion: The Power of a Strong Online Presence

These case studies illustrate how small businesses in various industries—services, e-commerce, and retail—have successfully leveraged their websites to grow their business, improve customer experience, and increase revenue. Whether through e-commerce integration, online booking systems, or a modern, mobile-friendly design, these businesses were able to overcome challenges and thrive in the digital age.

Future Trends in Web Design and Digital Marketing

As technology evolves and consumer expectations shift, it's important for small business owners to stay updated on the latest trends in web design and digital marketing. The online landscape is constantly changing, and businesses that can adapt quickly are more likely to maintain a competitive edge and continue attracting customers. In this section, we'll explore some of the most impactful future trends that are reshaping the way websites are built, marketed, and experienced by users.

Voice Search Optimization

With the rise of voice-activated assistants like Amazon Alexa, Google Assistant, and Apple's Siri, voice search is becoming a significant trend in how people interact with the internet. Consumers are increasingly using voice commands to search for products, services, and information, meaning businesses need to optimize their websites to accommodate this shift. How Voice Search Is Changing SEO:

Conversational Keywords: Voice search queries tend to be longer and more conversational compared to traditional text searches. For example, instead of typing "best pizza in Chicago," a voice search might be "What's the best place to get pizza in Chicago?" Optimizing for long-tail, conversational keywords will help your website rank better for voice searches.

Local SEO Focus: Voice searches are often location-based, with users asking things like "Where is the nearest coffee shop?" or "Plumbers near me." Ensuring that your website is optimized for local search terms and that your Google My Business profile is updated can help you rank higher in voice search results.

Featured Snippets: Search engines often pull voice search results from featured snippets—the small blocks of information that appear at the top of Google's search results. Structuring your content to answer common questions clearly and concisely can increase your chances of being featured.

What You Can Do:

Create FAQ pages that directly answer common questions your customers might ask through voice search.

Focus on conversational language and natural phrasing in your content.

Ensure your website is optimized for local SEO, with up-to-date contact details, business hours, and service areas.

Mobile-First Design and User Experience

With mobile devices accounting for more than half of global web traffic, mobile-first design is no longer optional—it's essential. Mobile-first means designing your website with mobile users as the primary audience, ensuring that your site works perfectly on smaller screens before optimizing for desktop.

Why Mobile-First Matters:

Google's Mobile-First Indexing: Google now primarily uses the mobile version of websites to determine search rankings. This means if your mobile site isn't optimized, your rankings could suffer, even on desktop searches.

User Expectations: Mobile users expect websites to load quickly, be easy to navigate, and provide a seamless browsing experience. A poor mobile experience can lead to high bounce rates, lost customers, and lower search rankings.

What You Can Do:

Responsive Design: Ensure that your website uses responsive design, which automatically adjusts the layout, images, and text based on the device being used.

Fast Load Times: Mobile users expect fast loading times. Compress images, enable browser caching, and use lightweight code to improve page speed.

Simplified Navigation: Streamline your site's navigation to make it easy for mobile users to find the information they need with minimal scrolling and tapping.

Artificial Intelligence (AI) and Chatbots

Artificial Intelligence (AI) is transforming how businesses interact with customers online. AI-powered chatbots are becoming more common, providing instant, 24/7 customer support and helping businesses handle inquiries more efficiently. AI can also be used for tasks like personalized content recommendations, automated customer service, and data analysis.

How AI and Chatbots Are Changing Web Design:

24/7 Customer Service: Chatbots allow businesses to engage with customers at any time, answering common questions, providing product recommendations, and even assisting with bookings or purchases.

Personalization: AI can analyze user behavior to deliver personalized experiences, such as recommending products based on past purchases or tailoring content to individual preferences.

Data-Driven Insights: AI can collect and analyze website data to help business owners make smarter decisions about design, content, and marketing strategies.

What You Can Do:

Consider integrating an AI-powered chatbot on your website to handle customer inquiries, especially if your business frequently deals with repetitive questions (e.g., operating hours, service availability, or product details).

Use AI tools to personalize the user experience on your website, such as recommending products based on browsing history or tailoring content for repeat visitors.

Leverage AI analytics tools to gather insights into customer behavior, helping you refine your marketing strategy and website layout.

Video Content and Visual Storytelling

Video content continues to be one of the most engaging forms of media online. As attention spans shorten, users are increasingly drawn to short, impactful videos that quickly convey information. Businesses that incorporate video content into their websites can improve engagement, boost conversion rates, and enhance their SEO rankings.

Why Video Content Is Critical:

Higher Engagement: Studies show that users spend more time on websites with video content. Whether it's a product demo, customer testimonial, or behind-the-scenes footage, videos capture attention more effectively than text alone.

SEO Benefits: Google prioritizes websites that use video content, as videos often lead to longer session duration and lower bounce rates—both positive signals for search engine algorithms.

Mobile Compatibility: Videos are especially effective on mobile devices, where users often prefer watching over reading.

What You Can Do:

Add product demo videos or explanation videos to key pages on your website to help potential customers better understand your offerings.

Use social proof by incorporating customer testimonials in video format, which can be more persuasive than written reviews.

Optimize your videos for SEO by including relevant keywords in the video titles, descriptions, and tags. Use YouTube or Vimeo to host your videos and embed them on your site to avoid slowing down page load times.

Personalization and Dynamic Content

Today's consumers expect personalized experiences, and dynamic content allows websites to adapt based on a visitor's behavior, preferences, or demographics. Personalization can significantly improve the user experience, resulting in higher engagement and increased conversions.

How Dynamic Content Works:

Tailored Experiences: Dynamic content changes based on user behavior, such as showing recommended products or personalized offers based on past purchases or browsing history.

Geo-Targeting: Websites can use the visitor's location to show region-specific content or promotions. For example, a clothing retailer might highlight winter coats for visitors in cold climates while showcasing swimsuits for those in warmer areas.

Email Personalization: When integrated with email marketing systems, dynamic content can be used to send personalized offers, product recommendations, or reminders based on a customer's previous interactions with your website.

What You Can Do:

Implement personalized product recommendations on your website's homepage or product pages, based on a user's past behavior.

Use geo-targeting to display location-based promotions or content that speaks directly to a visitor's region, increasing the relevancy of your offers.

Leverage email marketing tools that integrate with your website's data to send personalized emails that recommend products or offer discounts based on a customer's interests.

Augmented Reality (AR) and Immersive Experiences

Augmented Reality (AR) is gaining traction in web design, particularly for e-commerce and retail businesses. AR allows users to interact with products in a more immersive way, such as virtually trying on clothing or seeing how furniture would look in their home.

Why AR Is Important for the Future:

Improved User Experience: AR can provide a more engaging and interactive experience for users, helping them visualize products more clearly before making a purchase.

Reduced Returns: For e-commerce businesses, AR can reduce return rates by allowing customers to "try before they buy," ensuring the product fits their needs and expectations.

Competitive Advantage: As AR becomes more common, businesses that implement AR features on their websites will stand out from competitors and offer a cutting-edge experience.

What You Can Do:

If you sell physical products, consider integrating AR tools that allow customers to visualize how your products would look in real life. For example, a home decor store could use AR to let customers see how furniture or artwork would look in their space.

Use AR for virtual try-ons in fashion and beauty, allowing customers to see how clothing, makeup, or accessories would look before purchasing.

Sustainability and Green Hosting

As consumers become more environmentally conscious, sustainability is becoming a bigger focus in web design and hosting. Businesses are increasingly seeking out green hosting providers that use renewable energy to power their servers. Additionally, consumers are favoring brands that demonstrate eco-friendly practices, both online and offline.

Why Sustainability Matters:

Consumer Preferences: Many consumers prefer to support businesses that are committed to sustainability, so highlighting your eco-friendly practices can improve your brand's reputation and attract more customers.

Green Hosting: By choosing a green hosting provider, your website can reduce its carbon footprint. These hosting providers use renewable energy sources and implement energy-efficient technologies to minimize environmental impact.

What You Can Do:

Partner with a green hosting provider that powers its servers with renewable energy. Some hosting providers offer eco-friendly hosting options for businesses that want to minimize their environmental footprint.

Highlight your sustainability efforts on your website, such as using recyclable packaging, reducing energy consumption, or supporting environmental causes. This can resonate with eco-conscious customers and differentiate your brand.

Conclusion: Preparing for the Future of Web Design and Digital Marketing

The digital landscape is always evolving, and staying ahead of these trends can help your business remain competitive and relevant. From voice search optimization and AI-powered chatbots to personalized experiences and AR, these trends offer exciting opportunities to enhance your website and create more engaging experiences for your customers.

By staying informed about future trends and adapting your website accordingly, you'll be well-positioned to grow your online presence and continue meeting your customers' changing needs.

Bonus Resources: Tools and Platforms

Building and maintaining a successful website requires more than just a good design and basic content. There are a wide variety of tools and platforms available that can help you optimize your website, improve user experience, and grow your online business more efficiently. Whether you're looking to boost your search engine rankings, create stunning content, or streamline your e-commerce operations, the right tools can save you time, reduce costs, and enhance your website's performance.

In this chapter, we'll explore a range of essential tools across different categories, each of which can help small business owners build, manage, and grow their online presence.

SEO (Search Engine Optimization) Tools

Effective SEO is critical for ensuring that your website ranks well in search engines like Google, helping potential customers find you online. Below are some powerful tools that can help you optimize your website for search engines.

1. Google Search Console

What It Does: Google Search Console is a free tool from Google that helps you monitor and troubleshoot your website's presence in Google search results.

Key Features:

Performance Reports: See how your website is performing in search, including total impressions, clicks, and average search position.

Index Coverage: Check which pages of your site are indexed by Google and identify any errors or issues preventing your site from being fully indexed.

Mobile Usability: Monitor how mobile-friendly your website is and get recommendations for improving usability.

Why It's Useful: Google Search Console provides direct insights from Google, helping you identify and fix SEO issues, improve your site's visibility, and track your performance in search results.

2. SEMrush

What It Does: SEMrush is a comprehensive SEO and digital marketing tool that helps you track your website's SEO performance, analyze competitors, and identify new keyword opportunities.

Key Features:

Keyword Research: Find the best keywords to target for your SEO campaigns based on search volume, competition, and trends.

Competitor Analysis: Discover what keywords your competitors are ranking for and analyze their backlink strategies.

Site Audits: Run detailed site audits to identify technical SEO issues and get recommendations for fixing them.

Why It's Useful: SEMrush is a powerful all-in-one tool for managing SEO, tracking rankings, and analyzing competitors. It's particularly useful for small businesses looking to improve their search visibility and optimize their content.

3. Yoast SEO (for WordPress)

What It Does: Yoast SEO is a popular WordPress plugin that helps you optimize your website's on-page SEO, ensuring your content is structured in a way that search engines can understand.

Key Features:

On-Page Optimization: Get real-time suggestions for improving your content's SEO, including keyword placement, meta descriptions, headings, and readability.

XML Sitemaps: Automatically generate an XML sitemap to help search engines index your website more efficiently.

Readability Scores: Ensure your content is easy to read by checking sentence length, paragraph structure, and overall readability.

Why It's Useful: Yoast SEO simplifies the SEO process for WordPress users, making it easy to optimize each page and post without needing advanced technical knowledge.

4. Ahrefs

What It Does: Ahrefs is an advanced SEO tool that specializes in backlink analysis, keyword research, and competitive analysis.

Key Features:

Backlink Profile: Analyze your website's backlink profile and identify opportunities to build high-quality backlinks.

Keyword Explorer: Research keywords and see how difficult it is to rank for them, along with suggestions for related terms.

Content Gap Analysis: Discover which keywords your competitors are ranking for that you are not, helping you target new content ideas.

Why It's Useful: Ahrefs is excellent for businesses looking to boost their off-page SEO, build backlinks, and identify gaps in their content strategy.

Content Creation Tools

Compelling content is the cornerstone of any successful website. Whether you need high-quality images, engaging blog posts, or eyecatching graphics, the following tools can help you create and manage content efficiently.

1. Canva

What It Does: Canva is an easy-to-use graphic design tool that allows you to create professional-quality visuals for your website, social media, and marketing materials without needing advanced design skills.

Key Features:

Templates: Access thousands of customizable templates for social media posts, blog headers, infographics, and more.

Drag-and-Drop Editor: Create custom designs by simply dragging and dropping elements like text, images, icons, and shapes.

Free Stock Photos and Icons: Canva offers a large library of free and premium stock images, icons, and illustrations to enhance your designs.

Why It's Useful: Canva is ideal for small business owners who need high-quality visuals but don't have the time or budget to hire a professional designer.

2. Grammarly

What It Does: Grammarly is a writing assistant that helps you improve your grammar, spelling, punctuation, and overall writing style. It's perfect for creating polished, professional content for your website or blog.

Key Features:

Real-Time Grammar and Spell Check: Automatically corrects grammar, spelling, and punctuation errors as you write.

Style Suggestions: Provides recommendations to improve sentence structure, word choice, and clarity.

Plagiarism Checker: Checks for plagiarism to ensure your content is original.

Why It's Useful: Grammarly helps you produce error-free, high quality written content, which is crucial for maintaining a professional image and improving SEO.

3. BuzzSumo

What It Does: BuzzSumo is a content research tool that helps you discover trending topics and content ideas based on what's performing well in your industry.

Key Features:

Content Analysis: Analyze the most shared and engaged content across social media platforms, helping you identify popular topics.

Competitor Monitoring: Track your competitors' content performance and see what topics are resonating with their audience.

Influencer Research: Find influencers in your industry who can help you promote your content.

Why It's Useful: BuzzSumo is a great tool for developing content strategies, identifying trending topics, and staying ahead of your competitors in terms of content creation.

4. Hemingway App

What It Does: Hemingway App is a writing tool that focuses on improving the readability of your content by simplifying your sentences and making your writing clearer and more concise.

Key Features:

Readability Score: Provides a readability score based on sentence complexity, passive voice usage, and overall clarity.

Highlight Problem Areas: Highlights difficult-to-read sentences, excessive adverbs, and other areas for improvement.

Instant Feedback: Offers real-time feedback as you write or edit your content.

Why It's Useful: Hemingway App is perfect for businesses looking to create clean, readable content that engages visitors and encourages them to stay on your website longer.

E-Commerce Platforms

If you're planning to sell products or services online, an e-commerce platform is essential. Below are some of the best e-commerce tools to help you set up an online store, manage inventory, and streamline the checkout process.

1. Shopify

What It Does: Shopify is a leading e-commerce platform that allows you to build and manage an online store with ease. It's a fully hosted solution, meaning all you need is an account to start selling.

Key Features:

Drag-and-Drop Store Builder: Easily create a professional online store without coding.

Payment Integration: Accept payments via credit cards, PayPal, and other popular payment methods.

Inventory Management: Track stock levels and automatically update inventory as products are sold.

Built-In SEO Tools: Optimize product pages for search engines to increase visibility and attract more customers.

Why It's Useful: Shopify is an all-in-one solution for small businesses that want to sell products online without dealing with technical complexities. Its ease of use and scalability make it perfect for businesses of all sizes.

2. WooCommerce (for WordPress)

What It Does: WooCommerce is a free, open-source e-commerce plugin for WordPress that allows you to turn your WordPress website into a fully functional online store.

Key Features:

Customizable Storefront: Build a customizable online store that integrates seamlessly with your WordPress site.

Product Management: Add, manage, and organize products with ease, including digital downloads, physical products, or services.

Extensions and Integrations: WooCommerce offers a wide range of extensions for payment gateways, shipping options, marketing, and more.

Secure Payments: Accept secure payments through multiple payment gateways, including Stripe, PayPal, and others.

Why It's Useful: WooCommerce is an excellent choice for businesses that already use WordPress and want a flexible, customizable ecommerce solution.

3. BigCommerce

What It Does: BigCommerce is a powerful e-commerce platform designed for growing businesses that need advanced features and scalability.

Key Features:

Customizable Templates: Choose from a range of professional templates to build a modern, mobile-friendly online store.

Multi-Channel Selling: Sell on multiple platforms, including Facebook, Instagram, Amazon, and eBay, all from one dashboard.

Advanced SEO: Take advantage of BigCommerce's robust SEO features, including optimized URLs, automatic sitemaps, and metadata management.

Inventory and Order Management: Manage your entire inventory, process orders, and handle shipping logistics all from the BigCommerce dashboard.

Why It's Useful: BigCommerce is ideal for businesses that plan to scale their e-commerce operations and need more robust features than what basic platforms offer.

Email Marketing Tools

Email marketing remains one of the most effective ways to engage with customers, promote products, and drive traffic back to your website. Here are some top email marketing tools to help you build, manage, and automate your campaigns.

1. Mailchimp

What It Does: Mailchimp is one of the most popular email marketing platforms, offering easy-to-use tools for building email lists, creating campaigns, and tracking results.

Key Features:

Drag-and-Drop Email Builder: Design professional-looking emails with customizable templates and a drag-and-drop editor.

Automated Campaigns: Set up automated email sequences, such as welcome emails, abandoned cart reminders, and post-purchase follow-ups.

Analytics: Track email open rates, click-through rates, and subscriber activity to measure the effectiveness of your campaigns.

Segmentation: Segment your email list based on customer behavior, preferences, and demographics for more targeted campaigns.

Why It's Useful: Mailchimp's user-friendly interface and robust features make it an excellent choice for small businesses looking to grow and engage their audience through email marketing.

2. Constant Contact

What It Does: Constant Contact is a powerful email marketing platform known for its ease of use and comprehensive feature set for small businesses.

Key Features:

Email Templates: Choose from a wide variety of email templates designed for different industries and purposes.

List Management: Easily import, manage, and segment your email contacts for more personalized messaging.

Event Marketing: Create and promote events through email invitations, registration forms, and follow-up emails.

Reporting: Gain insights into your email campaign performance with real-time analytics, including open rates, click-through rates, and engagement.

Why It's Useful: Constant Contact is perfect for small businesses that need a simple, effective platform for building and managing email marketing campaigns.

3. Sendinblue

What It Does: Sendinblue is an all-in-one marketing platform that combines email marketing with SMS marketing, CRM, and automation tools.

Key Features:

Email and SMS Campaigns: Send both email and SMS campaigns from the same platform.

Automation Workflows: Create automated workflows for customer journeys, such as welcome sequences or post-purchase follow-ups.

Personalization: Personalize emails with customer data, ensuring your messages are relevant and engaging.

CRM Integration: Manage customer relationships and keep track of interactions with built-in CRM tools.

Why It's Useful: Sendinblue's combination of email, SMS, and CRM makes it a great choice for businesses looking to create more integrated, multi-channel marketing campaigns.

Website Analytic Tools

Understanding how visitors interact with your website is crucial for improving user experience, optimizing content, and increasing conversions. Here are some of the top analytics tools that can help you track and analyze your website's performance.

1. Google Analytics

What It Does: Google Analytics is the industry standard for tracking website traffic and user behavior. It provides detailed insights into how visitors find your site, which pages they visit, and how they interact with your content.

Key Features:

Traffic Insights: See where your traffic is coming from, whether it's search engines, social media, or direct visits.

User Behavior: Track user behavior on your site, including how long they stay on each page, what they click on, and when they leave.

Goal Tracking: Set up goals to track specific actions, such as purchases, form submissions, or downloads.

Audience Reports: Understand your audience's demographics, interests, and location.

Why It's Useful: Google Analytics provides a wealth of data that can help you make informed decisions about your website's design, content, and marketing strategy.

2. Hotjar

What It Does: Hotjar is a behavior analytics tool that helps you visualize how users interact with your website through heatmaps, session recordings, and feedback surveys.

Key Features:

Heatmaps: See where users are clicking, scrolling, and hovering on your website, helping you identify high-engagement areas.

Session Recordings: Watch recordings of real user sessions to see how visitors navigate your site and identify usability issues.

Surveys and Feedback Forms: Collect direct feedback from users to understand their experience and identify areas for improvement.

Why It's Useful: Hotjar is an excellent tool for optimizing user experience by providing visual insights into how visitors interact with your website.

3. Crazy Egg

What It Does: Crazy Egg is a website optimization tool that uses heatmaps, scroll maps, and A/B testing to help you understand user behavior and improve your website's performance.

Key Features:

Heatmaps and Scroll Maps: Visualize how far down users scroll on your pages and where they're clicking.

A/B Testing: Test different versions of your website's pages to see which design or layout performs better.

User Recordings: Watch real user sessions to identify pain points and improve your site's navigation.

Why It's Useful: Crazy Egg helps you fine-tune your website's design and user experience by providing actionable insights based on user behavior.

Conclusion: Empowering Your Website with the Right Tools

The tools and platforms outlined in this chapter provide small business owners with the resources needed to build, optimize, and grow their websites efficiently. Whether you're looking to improve SEO, create engaging content, streamline e-commerce operations, or track website performance, these tools will help you achieve your goals more effectively.

By leveraging the right combination of tools, you can enhance your online presence, attract more visitors, and ultimately, convert those visitors into loyal customers. With solutions ranging from SEO and content creation to email marketing and analytics, you'll be well equipped to manage your website like a pro and drive sustained growth for your business.

Small Business Website Checklist

1. Planning and Preparation

Before you start building your website, it's essential to plan and outline your goals, audience, and key features.

- **Define Website Goals:**

What do you want your website to achieve? (e.g., attract new customers, sell products, showcase services)

- **Identify Target Audience:**

Who are your customers? What are their needs, and how can your website meet them?

- **Decide on Key Features:**

Will you need an e-commerce store, booking system, blog, or portfolio?

- **Choose Your Website Type:**

Will you use a DIY platform (e.g., Wix, WordPress) or hire a professional designer?

- **Budget for the Website:**

Determine your budget for design, development, hosting, and ongoing maintenance.

2. Domain and Hosting Setup

Your domain name and hosting are the foundation of your website. This section covers everything needed to get your site online.

- **Choose a Domain Name:**

Pick a simple, memorable, and relevant domain name (e.g., yourbusiness.com).

- **Register Your Domain:**

Use a domain registrar or your hosting provider to secure your chosen domain name.

- **Select a Web Hosting Plan:**

Choose a hosting plan that suits your needs (shared hosting, VPS, dedicated hosting). Many hosting providers offer flexible hosting options for small businesses.

- **Set Up SSL Certificate:**

Install an SSL certificate to secure your site and enable HTTPS for data encryption and better SEO.

3. Website Design and Development

Designing a professional, user-friendly, and functional website is key to engaging visitors and converting them into customers.

- **Choose a Website Platform:**

Decide on a platform (e.g., WordPress, Wix, Shopify) or hire a web developer if you need custom features.

- **Select a Mobile-Responsive Design:**

Ensure your website design is mobile-friendly and works well on all devices.

- **Create a Sitemap and Structure:**

Plan your website's structure, including key pages (e.g., homepage, about, services, products, contact).

- **Design Branding Elements:**

Create a cohesive brand identity, including logo, color scheme, and fonts that align with your business.

- **Develop Key Pages:**

Write copy and add content for all core pages (homepage, services, contact page, etc.).

- **Add Visuals:**

Use high-quality images and videos that represent your business and products effectively.

- **Set Up Calls-to-Action (CTAs):**

Ensure clear CTAs are placed throughout the site (e.g., "Contact Us," "Shop Now," "Get a Free Quote").

- **Integrate Key Features:**

Set up features such as e-commerce (if applicable), appointment booking, and contact forms.

4. Content Creation

Your content plays a major role in engaging visitors and driving traffic. It needs to be well-written, SEO-friendly, and relevant to your audience.

- **Create a Content Strategy:**

Plan what content your website will need, such as service pages, blogs, product descriptions, and FAQs.

- **Write SEO-Optimized Copy:**

Use relevant keywords and make sure content is clear, informative, and optimized for search engines.

- **Develop Blog or Resource Pages:**

Create educational or engaging blog posts to drive traffic and improve SEO.

- **Ensure Readability:**

Use tools like Grammarly or Hemingway to ensure content is errorfree and easy to read.

- **Include Visual Content:**

Add infographics, product images, videos, and other visual elements that enhance the user experience.

5. SEO and Analytics Setup

To make sure your website ranks well in search engines and attracts visitors, SEO and performance tracking should be top priorities.

- **Conduct Keyword Research:**

Identify the best keywords for your business using tools like Google Keyword Planner, SEMrush, or Ahrefs.

- **Optimize Meta Tags:**

Write SEO-friendly title tags and meta descriptions for each page.

- **Set Up Google Analytics:**

Install Google Analytics to track website traffic, user behavior, and conversions.

- **Set Up Google Search Console:**

Register your site with Google Search Console to monitor your site's indexing status and SEO performance.

- **Create an XML Sitemap:**

Generate an XML sitemap and submit it to Google Search Console to ensure all your pages are indexed.

- **Optimize for Local SEO:**

Set up your Google My Business profile and include your business's address and contact information for local searches.

- **Create Backlinks:**

Build backlinks by connecting with reputable websites in your industry or through guest blogging.

6. Testing and Launch

Before going live, thoroughly test your website to ensure everything functions properly and provides a smooth user experience.

- **Test All Functionality:**

Check that all contact forms, shopping carts, booking systems, and links are working correctly.

- **Check Mobile Responsiveness:**

Test your site on various mobile devices and browsers to ensure it displays correctly.

- **Check Page Load Speed:**

Use tools like Google PageSpeed Insights or GTmetrix to test and improve your site's load time.

- **Cross-Browser Testing:**

Test your site in different browsers (Chrome, Firefox, Safari) to ensure compatibility.

- **Proofread Content:**

Review all text on the site to catch any typos or errors.

- **Check for SEO Issues:**

Run an SEO audit to identify and fix any missing metadata, broken links, or duplicate content issues.

7. Post-Launch and Ongoing Maintenance

Once your site is live, ongoing maintenance and marketing are crucial to keeping it running smoothly and attracting new visitors.

- **Submit Website to Search Engines:**

Ensure your website is submitted to Google and Bing for indexing.

- **Set Up Regular Backups:**

Schedule regular backups of your site to prevent data loss in case of issues or hacks.

- **Perform Regular Software Updates:**

Update your CMS, plugins, and themes regularly to keep your site secure.

- **Monitor Website Performance:**

Use Google Analytics and other tools to track visitor behavior, bounce rates, and conversions.

- **Add New Content Regularly:**

Update your site with new blog posts, news, or product updates to keep visitors engaged and improve SEO.

- **Check Security Regularly:**

Run security scans to ensure your site is protected from malware, and consider adding a firewall for extra protection.

- **Engage in Digital Marketing:**

Run email marketing campaigns, social media ads, or PPC (Pay-PerClick) campaigns to drive traffic to your site.

- **Track SEO Performance:**

Regularly check search rankings and tweak your SEO strategy based on keyword performance and traffic trends.

8. Growth and Scalability

As your business grows, you'll want to ensure that your website can handle increased traffic and new features.

- **Upgrade Hosting as Needed:**

As traffic increases, consider upgrading from shared hosting to VPS or dedicated hosting.

- **Add E-Commerce Capabilities (if applicable):**

If you haven't already, integrate e-commerce features like a shopping cart and payment gateway.

- **Expand Features:**

Consider adding customer portals, appointment booking, or new sections for additional services or products.

- **Track Growth with KPIs:**

Keep monitoring key performance indicators like traffic growth, conversion rates, and customer acquisition costs to ensure your website is supporting your business's growth goals.

Setting Your Website Up for Long-Term Success

This checklist provides a clear roadmap for planning, building, and maintaining your small business website. By following each step, you'll be well on your way to launching a successful online presence that attracts visitors, converts leads, and helps your business grow. Whether you're starting with a simple brochure site or building a full e-commerce platform, taking the time to properly plan, optimize, and maintain your website will pay off in the long run.

Frequentlly Asked Questions (FAQs)

1. How much does it cost to build a small business website?

The cost of building a website depends on several factors, including the complexity of the site, the features you need, and whether you build it yourself or hire a professional. Here's a breakdown of potential costs:

DIY Website Builders: Platforms like Wix, Squarespace, or WordPress can cost anywhere from $10 to $50 per month for basic plans, depending on the features you need (e.g., e-commerce, custom domain, SEO tools).

Hiring a Professional Designer/Agency: If you hire a web designer or agency, the costs will be higher. A simple website can start at around $1,000 to $5,000, while more complex websites with custom features can range from $5,000 to $20,000 or more.

Additional Costs: Don't forget ongoing costs like domain registration (typically $10 to $50 per year), hosting (from $5 to $100+ per month), and website maintenance or updates.

Pro Tip: If you're just starting out, it's okay to begin with a simple, budget-friendly solution and upgrade as your business grows. Many hosting companies offer affordable hosting and design packages tailored to small business needs.

2. How long does it take to build a website?

The time it takes to build a website depends on the complexity of the site and your chosen method:

DIY Platforms: If you're building your website yourself with a DIY platform, it can take anywhere from a few days to a couple of weeks, depending on how quickly you can add content and customize the design.

Hiring a Professional: If you're working with a professional designer or agency, the timeline will vary based on the scope of the project. A simple, brochure-style website can take 2-4 weeks, while more complex sites with custom features or e-commerce capabilities can take 6-12 weeks or longer.

To speed up the process, have all your content (text, images, product descriptions, etc.) ready before the project begins.

3. Do I need to know how to code to create a website?

No, you don't need to know how to code to create a website. Many platforms, like Wix, Squarespace, and WordPress, offer drag-anddrop builders that allow you to design and customize your site without any technical knowledge.

However, if you need a more complex or custom-built site (e.g., for e-commerce or with advanced features), it might be worth hiring a web designer or developer. Many hosting companies offer both DIY website building tools and professional design services, depending on your needs.

4. What is hosting, and why do I need it?

Hosting is the service that allows your website to be stored on the internet and accessible to visitors. Essentially, your website's files (like text, images, and videos) are stored on a server provided by a hosting company, which makes your site accessible via your domain name.

Without hosting, your website wouldn't be live on the web. There are several types of hosting available:

Shared Hosting: The most affordable option, where multiple websites share the same server.

VPS Hosting: Offers more resources and flexibility than shared hosting, suitable for growing businesses.

Dedicated Hosting: A more expensive option that gives your website its own server, ideal for large websites with high traffic.

Many hosting providers offers various hosting packages, from shared hosting for small businesses to more advanced options like VPS and dedicated hosting as your website grows.

5. How do I choose the right domain name?

Your domain name is your website's address (e.g., www.
yourbusiness.
com), and choosing the right one is important for branding and SEO. Here are some tips to help you choose:

Keep It Simple and Memorable: A short, easy-to-spell domain name is easier for customers to remember.

Use Keywords (If Relevant): If possible, include a keyword that describes your business or services. For example, "www.MiamiPlumbing.com" clearly tells visitors what the business does and where it operates.

Avoid Numbers and Hyphens: These can be confusing and harder for people to remember.

Choose the Right Extension: .com is the most common and trusted, but if it's unavailable, consider alternatives like .net, .org, or industry specific options like .shop or .tech.

Hosting and development companies, can help you secure the perfect domain name for your business, with domain registration services included in many of our hosting plans.

6. Do I need SEO, and how do I get started?

Yes, Search Engine Optimization (SEO) is essential if you want your website to be found on search engines like Google. SEO helps improve your website's visibility in search results, which can drive more organic (unpaid) traffic to your site. Here are the basic steps to get started with SEO:

Keyword Research: Identify the search terms potential customers are using to find businesses like yours (e.g., "custom cakes in Chicago").

On-Page Optimization: Use these keywords naturally in your website's content, page titles, meta descriptions, and headings.

Mobile-Friendly Design: Google prioritizes websites that work well on mobile devices, so make sure your website is responsive.

Backlinks: Earn backlinks from reputable websites to improve your site's authority in search engine rankings.

Some hosting companies offer SEO services to help small businesses optimize their websites and rank higher in search results.

7. How often should I update my website?

You should update your website regularly to keep it fresh, secure, and relevant to your audience. Here's a general guide:

Content Updates: Add new blog posts, product updates, or company news at least once a month. Regular updates signal to search engines that your website is active and current.

Security Updates: If you're using a content management system (CMS) like WordPress, it's important to update your software, plugins, and themes regularly to prevent security vulnerabilities. Design Refreshes: Every 1-2 years, consider reviewing your site's design to ensure it still looks modern and meets your business's needs.

Some hosting providers offer website maintenance packages that include regular updates, backups, and security checks.

8. How do I know if my website is successful?

The success of your website depends on your business goals, but here are some key metrics to track:

Traffic: How many people are visiting your website? Tools like Google Analytics can help you track the number of visitors, where they're coming from, and which pages they're viewing.

Bounce Rate: This shows the percentage of visitors who leave your site after viewing just one page. A high bounce rate may indicate that your content isn't engaging enough or that your site needs improvement.

Conversion Rate: How many visitors are taking the desired action on your site (e.g., making a purchase, submitting a contact form)? A low conversion rate may indicate that you need to optimize your CTAs (calls to action) or checkout process.

Engagement: How long are visitors staying on your site, and how many pages are they viewing? The longer they stay and the more pages they visit, the better your engagement levels.

Some hosting companies provide analytics and performance tracking tools to help you measure the success of your website and identify areas for improvement.

9. What do I do if my website is hacked?

Website hacking is a serious issue, but if your site is compromised, here's what you should do:

Take Your Website Offline: If you suspect your site has been hacked, immediately take it offline or put it into maintenance mode to prevent further damage.

Restore from Backup: If you have recent backups of your website, restore the site to a clean version before the hack occurred.

Change Passwords: Reset all admin passwords and update your security settings to prevent further breaches.

Check for Malware: Run a malware scan and remove any malicious code that has been added to your website.

10. What are the best ways to promote my website once it's live?

Once your website is live, you'll need to promote it to attract visitors. Here are some effective strategies:

SEO (Search Engine Optimization): Optimize your site's content to rank higher on Google for relevant search terms.

Social Media Marketing: Share your website on platforms like Facebook, Instagram, and LinkedIn. Create content that encourages users to visit your site, such as blog posts, promotions, or behind-the-scenes videos.

Email Marketing: Build an email list and send regular newsletters or updates that drive traffic back to your website.

PPC (Pay-Per-Click) Advertising: Run Google Ads or social media ads to reach more potential customers. PPC ads can target specific demographics or keywords related to your business.

A Message from the Authors

As a small business owner, you already know that running a successful business in today's fast-paced and competitive economy comes with many challenges. Building and managing your website is one of those challenges, and while it's possible to handle some aspects yourself, the truth is that for real efficiency, quality, and long term results, working with a professional web hosting and development team can make all the difference.

Why? Because today's websites are much more than just digital brochures—they are powerful business tools that require expertise in multiple areas to be truly successful. Even a basic website needs to be fast, secure, optimized for search engines (SEO), and designed to grow with your business. That's not to mention the potential need for integrations, automations, scheduling systems, AI, and interactive forms that allow you to better serve and communicate with your customers.

A professional, managed web hosting and development service can help you navigate all these complexities, allowing you to focus on what you do best—running your business. For as little as $100 to $300 per month, even the smallest business can invest in a managed service that handles all the critical aspects of your online presence— from fast, reliable hosting to seamless integrations and ongoing development support.

At BrevardHost.net, we specialize in providing fully managed web hosting and development services designed specifically for small businesses like yours. We take care of everything, from SEO and performance optimization to development and security, so you can be confident that your website is always running smoothly and effectively.

If your local market doesn't offer fully managed web hosting and development, or if you're simply ready to take your website to the next level, we're here to help. Visit us at https://brevardhost.net or feel free to contact us directly at authors@diakakis.us. We look forward to partnering with you and helping your business thrive online.

Glossary

This glossary provides quick definitions of common terms you'll encounter while managing and growing your small business website. Understanding these key concepts will help you navigate the technical aspects of web hosting, development, and digital marketing with confidence. Whether you're optimizing for SEO, improving user experience, or analyzing your website's performance, this glossary serves as a handy reference guide for your online journey.

A/B Testing

A method of comparing two versions of a web page or app to determine which one performs better. A/B testing helps improve conversion rates by testing different designs, calls-to-action, or content to see which variation drives more user engagement.

Alt Text

Short for "alternative text," alt text is a description of an image added to the HTML of a web page. This text helps search engines understand what the image is about and improves accessibility for visually impaired users who rely on screen readers.

Backlinks

Links from one website to another. Backlinks, also known as "inbound links," are a crucial factor in search engine rankings. Websites with many high-quality backlinks tend to rank higher in search results because they are seen as authoritative and trustworthy.

Bounce Rate

The percentage of visitors who land on a website and leave without interacting with any other pages. A high bounce rate may indicate that the content or user experience on the page needs improvement to keep visitors engaged.

Browser Cache

A temporary storage location on your computer where web pages are stored after being downloaded by the browser. Caching improves load times by allowing the browser to reuse stored files instead of downloading them again when revisiting a website.

Call to Action (CTA)

A button, link, or prompt that encourages website visitors to take a specific action, such as "Buy Now," "Sign Up," or "Contact Us." Effective CTAs are clear, action-oriented, and strategically placed to guide users through the conversion process.

CMS (Content Management System)

A software platform used to create, manage, and modify digital content on websites without requiring advanced coding knowledge. Examples include WordPress, Joomla, and Shopify. CMS platforms make it easy to update websites, add new content, and manage design elements.

Conversion Rate

The percentage of website visitors who complete a desired action, such as making a purchase, signing up for a newsletter, or submitting a contact form. A higher conversion rate indicates that a website is effectively converting visitors into customers or leads.

CSS (Cascading Style Sheets)

A style sheet language used to define the look and feel of a website, such as layout, colors, fonts, and spacing. CSS works with HTML to control the visual presentation of web pages and ensure that designs are consistent across different devices and browsers.

DNS (Domain Name System)

The system that translates human-friendly domain names (e.g., www.example.com) into IP addresses that computers use to identify each other on the internet. DNS servers act as a phonebook for the web, allowing users to reach websites using easy-to-remember domain names.

Domain Name

The address of your website that users type into a browser to visit your site (e.g., www.yourbusiness.com). Domain names are unique and must be registered through a domain registrar.

E-Commerce

The process of buying and selling goods and services online. Ecommerce websites allow businesses to sell products, accept payments, and manage orders via the internet. Popular ecommerce platforms include Shopify, WooCommerce, and BigCommerce.

Google Analytics

A free tool from Google that provides detailed insights into website traffic, user behavior, and marketing performance. Google Analytics allows you to track metrics such as page views, bounce rates, session duration, and conversion rates, helping you optimize your website for better results.

Hosting (Web Hosting)

A service that stores your website's files and makes them accessible to users on the internet. Hosting providers allocate server space for websites and provide the infrastructure needed for your site to be live and accessible via your domain name. Hosting options include shared hosting, VPS, and dedicated hosting.

HTML (HyperText Markup Language)

The standard coding language used to create and structure content on web pages. HTML defines the layout of a webpage by using a series of tags and elements, such as headings, paragraphs, images, and links.

HTTPS (HyperText Transfer Protocol Secure)

A secure version of HTTP, the protocol used to transfer data between a user's browser and the website's server. HTTPS encrypts this data, providing a more secure connection and protecting sensitive information, such as credit card details and passwords.

IP Address

A unique string of numbers that identifies each device connected to the internet. Websites have their own IP addresses, which are mapped to domain names through the DNS system.

Keyword

A word or phrase that describes the content of a web page and matches the search terms users type into search engines. Keywords are an essential component of SEO, as they help search engines determine what your website is about and how it should rank in search results.

Landing Page

A standalone web page created specifically for a marketing or advertising campaign. Its primary goal is to convert visitors by prompting them to take a specific action, such as signing up for a newsletter, making a purchase, or downloading a resource.

Meta Description

A brief summary of a web page's content that appears in search engine results under the page's title. Meta descriptions help users decide whether to click on a link and can influence search engine rankings.

Mobile-First Design

A design approach that prioritizes the mobile version of a website first and then scales up for larger screens, such as tablets and desktops. With mobile traffic surpassing desktop traffic, mobile-first design ensures that websites provide a seamless experience for mobile users.

On-Page SEO

The practice of optimizing individual web pages to improve search engine rankings and drive traffic. On-page SEO includes optimizing title tags, meta descriptions, content, internal links, and images to make the page more relevant and easier for search engines to index.

Organic Traffic

Visitors who arrive at your website through unpaid search results, rather than through paid advertisements. Organic traffic is generated through effective SEO and content marketing strategies.

PPC (Pay-Per-Click)

An online advertising model where advertisers pay a fee each time one of their ads is clicked. PPC ads typically appear at the top of search engine results pages or on social media platforms, and they target specific keywords or audiences to drive traffic to a website.

Responsive Design

A web design approach that ensures a website adjusts smoothly to different screen sizes and devices, from desktops to smartphones. Responsive design improves user experience and is also a ranking factor in search engine algorithms.

SEO (Search Engine Optimization)

The process of optimizing a website to improve its visibility on search engines like Google. SEO involves improving both on-page and off-page elements, including keywords, meta tags, site speed, and backlinks, to rank higher in search results and attract more organic traffic.

SSL Certificate

A digital certificate that encrypts the data transmitted between a website and a user's browser. Websites with SSL certificates display "HTTPS" in the URL and a padlock icon, which indicates a secure connection. SSL certificates are essential for e-commerce sites and any site that handles sensitive data.

Sitemap

A file that lists all the pages on your website, designed to help search engines crawl and index your site more effectively. XML sitemaps are particularly useful for larger websites with many pages.

Traffic Source

The origin of your website visitors. Common traffic sources include organic search (unpaid search engine results), direct traffic (users typing in your URL), referral traffic (from other websites), and social media. Understanding traffic sources helps you optimize your marketing efforts.

UI (User Interface)

The layout and design of a website or app that users interact with. Good UI design ensures that elements like buttons, menus, and forms are easy to use and navigate, enhancing the overall user experience.

UX (User Experience)

The overall experience a user has when interacting with a website or digital product. UX encompasses everything from the ease of navigation to the site's load time, visual appeal, and how well the website meets the user's needs.

VPS (Virtual Private Server)

A type of web hosting where a physical server is divided into multiple virtual servers, each with its own resources. VPS hosting offers more control and flexibility than shared hosting, making it suitable for growing businesses that need more bandwidth and security.

Widget

A small, standalone application or tool that can be added to a website's sidebar, footer, or other areas. Widgets often display dynamic content such as calendars, social media feeds, or weather updates.

Biographies

Dr. Julia Diakakis

Dr. Julia Diakakis, a highly respected and decorated educator and businesswoman, boasts an impressive career spanning over 35 years. Born into a military family in Bremerhaven, Germany, Dr. Diakakis was raised throughout Europe providing her a diverse upbringing which laid the foundation for a versatile professional journey.

A proud graduate of Ohio University, she holds a B.S. in Education with a minor in English. Her certifications span various subjects including English, Math, History, Social Science, Earth Space Science, and Elementary Education with endorsements in Reading and English Speakers of Other Languages (ESOL). Her passion for education led her to different roles in educational institutions ranging from elementary to high school levels, earning her recognition as one of Florida's most esteemed educators. Furthering her passion for education, she has served on dissertation committees as the content expert for doctoral students.

Parallel to her educational career, Dr. Diakakis has been maintaining an IT business since 1994 and served the Brevard School Board as an IT Specialist for seven years. This combination of education and technology expertise uniquely positions her at the intersection of these two critical fields. Dr. Diakakis further expanded her skill set by obtaining a master's degree in educational administration, leading to her principal certification and an administrative role for about 15 years. She didn't stop there; her boundless passion for education, business, and information technology propelled her to pursue a Doctorate in Organizational Leadership and Development.

A member of prestigious societies such as Alpha Chi Honor Society, Kappa Delta Pi Honor Society, and National Society of School Administrators, Dr. Diakakis has made significant contributions to developing computer lab programs for Brevard Public Schools in Florida. She also pioneered the first fully operational Discourse Study Computer classroom in Florida. Her extensive work with numerous private companies throughout her career testifies to her business acumen.

Dr. Diakakis' unique blend of educational knowledge and business insights makes her a formidable force in her field. Her life and career serve as an inspiration for many who aspire to make a difference in the realms of education and business.

Contact

E-mail: authors@diakakis.us

Websites:

https://diakakis.us

https://BrevardHost.net

https://alphaomegacom.net

Saki Diakakis

Born in Athens, Greece, Saki Diakakis embarked on his journey with a classic high school education followed by military service as an Amphibian Green Beret. In 1989, Saki immigrated to the United States to nurture his passion for aviation and began training as a private pilot, initially aiming for a career as a commercial pilot.
However, he found himself drawn towards the burgeoning field of computers, networks, and software development. Saki returned to academia, this time immersing himself in computer science, and subsequently ventured into diverse IT industries.

From the early 1990s, his career path led him to work with local city government bodies as a Network Systems Analyst and Software Developer. His professional portfolio expanded across multiple sectors including county, city, and state governments, in areas of advertising, space subcontractors, home builders, and water treatment plants, among others. During his time with city government he designed and built the TV station for the City of Cocoa Beach, marking a significant milestone in his career.

During a nine-year stint in the advertising industry, Saki managed central networks and developed integral systems for the national yellow page company. As a consultant, he contributed to data analysis and application development. Over the years, he has developed an array of commercial applications encompassing document management systems, CRM, billing, building department permits, fleet management, and more. Adding another feather to the cap, he pioneered one of the first surveillance wireless undercover applications in the United States.

For the past 12 years, Saki has been associated with a local Managed Services Provider, specializing in computer networks, telephony, and cybersecurity applications. His services have benefited hundreds of businesses on local, federal, and international levels.

Saki's extensive formal education and practical experience have equipped him with knowledge that many businesses and stakeholders consider invaluable for their success. He is committed to leveraging this expertise to continue the contribution to the field of technology and business.

Contact

E-mail: authors@diakakis.us

Websites:

https://diakakis.us

https://BrevardHost.net

https://alphaomegacom.net

NOTES:

NOTES:

NOTES:

Made in United States
Orlando, FL
04 November 2026

90201196R00099